Geo-participation

Shanqi Zhang

Geo-participation

Typology, Technology and Applications

Springer

Shanqi Zhang
School of Architecture and Urban Planning
Nanjing University
Nanjing, Jiangsu, China

ISBN 978-981-95-8227-3 ISBN 978-981-95-8228-0 (eBook)
https://doi.org/10.1007/978-981-95-8228-0

This Springer imprint is published by the registered company Springer Nature Singapore Pte Ltd.
The registered company address is: 152 Beach Road, #21-01/04 Gateway East, Singapore 189721, Singapore

Acknowledgment This book is based largely on research conducted during my Ph.D. studies and was further revised and refined after my graduation. I am deeply grateful to my Ph.D. supervisor Dr. Rob Feick, whose insightful guidance, intellectual rigor, and constant encouragement shaped this work at every stage.

I acknowledge GeoThink research project, which supports my Ph.D. research and provides me with the opportunity to connect with scholars, students, and research opportunities that have greatly helped my professional development. I am also grateful to the National Natural Science Foundation of China for providing the funding support that allows this research to be carried forward and ultimately developed into this book.

A debt of gratitude goes to KWAccessAbility, the City of Kitchener, and Kitchener's farmers' market for their assistance with recruiting citizen participants for my study. I thank all of the citizen participants and local government interviewees for their kindness in taking the time to participate, and for sharing their insights, without which this research could not have been completed.

Finally, I appreciate the many colleagues and friends who offered feedback, discussion, and encouragement at different stages of this project.

Competing Interests This study was funded by the National Natural Science Foundation of China (52478061). The fieldwork of this study was funded by the GeoThink project.

The authors have no conflicts of interests to declare.

Ethical Approval Ethical approval was granted by the Ethics Committee of the University of Waterloo.

Informed consent to participate was obtained from individual participants.

All data has been anonymized.

Contents

List of Figures

List of Tables

Chapter 1
Introduction

Abstract This chapter introduces the context and motivation of this book, focusing on how geospatial technologies play an increasingly important role in public participation. New forms of public participation have emerged that require not only new tools for making use of the technology but also critical examinations from both government and citizen perspectives. This book proposes a new typology for geo-participation. Based on several empirical studies, it also sheds light on how local governments may solicit citizen input through new channels and enable new forms of government–citizen interaction.

1.1 Context and Motivation

With over 50% of the world's population now living in cities and the number continuing to grow (United Nations, 2016), cities are increasingly responsible for delivering services to people and businesses as well as for meeting difficult social, economic, and environmental challenges associated with sustainable forms of development (Roche, 2014). Local governments have increasingly capitalized on technologies, Internet and Communication Technologies (ICTs) and Web 2.0 technology, in particular, to enhance connections with the public, promote participatory decision-making to better address public needs, and improve the efficiency of service delivery by collaborating with the public (McKinsey Center for Government, 2017; Sivarajah et al., 2015). Despite these efforts, governments are still struggling with the increasing expectations and demands of citizens (McKinsey Center for Government, 2017). Although the importance of improving citizen participation and engagement for policy and decision-making has been recognized for some time now (Layne & Lee, 2001; United Nations, 2008), a recent report from The United Nations (2016) suggests that the progress in practice has been modest. Citizen participation initiatives still struggle with low participation rates and the difficulty of integrating participation outcomes with government procedures (Anthopoulos et al., 2014).

Geospatial technologies have been used in public participation processes to encourage public engagement, identify community needs, and enhance participatory

S. Zhang, *Geo-participation*, https://doi.org/10.1007/978-981-95-8228-0_1

decision-making (Sawicki & Peterman, 1996; Tang & Liu, 2015). Public agencies (e.g., local governments) use spatial information and methods (e.g., mapping) to provide the public with access to spatial information (e.g., spatial background of projects, spatial data about local environments) and to collect local spatial knowledge from the public (Brown & Kyttä, 2014; Rinner, 2006). More recent developments in geospatial technologies, such as the Geoweb and location-aware devices and services, have provided people with easier access to geospatial information and tools and enabled wider-scale spatial data generation and collaboration (Sieber et al., 2016). Local governments now have the opportunity to obtain local spatial knowledge about more diverse topics from the broader public and to use this knowledge to improve their data, services, and decision-making. However, local governments still face significant challenges in adopting the practices of citizens generating geospatial information because of perceived risks associated with the uncertain quality of citizen-generated data, needs of technical, financial, and human resources, increased citizen requests and demands, and lack of regulation and legislation (Brandeis & Nyerges, 2016; Johnson & Sieber, 2012, 2013). It is necessary to examine the participation potential of citizens generating geospatial information in local government settings and to assess the benefits and challenges associated with these practices.

It is against this backdrop that this book is carried out to investigate opportunities and challenges for local governments to take advantage of advances in geospatial technologies to enhance public participation. The rest of this chapter details the concept of public participation and reviews progress in public participation practices, particularly those that adopt geospatial information and technologies, in local government settings.

1.2 Research Objectives and Scope of This Book

The Geoweb-related technologies present significant potential for local governments to improve public participation and embark on new participatory initiatives. Yet, this potential needs to be further examined and assessed. This book aims to provide a conceptual and empirical investigation of emerging geo-participation practices by addressing the following research objectives: (1) situating geo-participation in the local government context; (2) empirically examining the use of geosocial media as an example of geo-participation for local governments to understand public opinions and to communicate non-emergency issues with citizens.

In this book, I chose geosocial media for empirical investigations of geo-participation considering its participatory potential. Social media is defined as "a group of Internet-based applications that build on the ideological and technological foundations of Web 2.0, and that allow the creation and exchange of user-generated content" (Kaplan & Haenlein, 2010, p. 61). In addition to this definition, social media is characterized by providing a platform for Internet users to socialize and connect with each other and to achieve certain goals with others who share common goals or interests (Magro, 2012). Exemplar applications include social networking applications such as Facebook, microblogging services such as Twitter, content-sharing services such as blogs and wikis, and media-sharing sites such as YouTube and Instagram (Magro, 2012). Geosocial media refers to social media that is associated with implicit or explicit geographic information (e.g., geotags, coordinates, and place names).

The participatory potential of geosocial media is twofold. First, social media is a widely used communication and citizen engagement tool by municipal governments (Reddick & Norris, 2013). Social media is commonly used by local governments to inform citizens of local issues and new policies or redirect social media users to other government websites, articles, and news (Khan et al., 2014). It is also considered to be an important engagement tool for governments to enhance their connection and interaction with citizens as well as to increase participation rates (Lev-On & Steinfeld, 2015; López-Ornelas & Zaragoza, 2014). Recent studies also suggest the appealing potential of social media as a platform to facilitate government–citizen collaboration. For example, Linders (2012) suggests that social media provides an environment for governments and citizens to communicate real-time information and foster collaboration and innovation.

Second, geosocial media has potential as a data source for monitoring urban environments and understanding public needs (Poorthuis & Zook, 2017). A range of studies examined the usefulness of geosocial media in understanding dynamic environments and enhancing situation awareness (De Longueville et al., 2010; Spinsanti & Ostermann, 2013). Other studies have suggested the potential of geosocial media for understanding public perceptions in terms of collective sentiments of local environments and infrastructure (Cao et al., 2014; Mitchell et al., 2013).

Despite the potential, using geosocial media for participatory purposes reveals several issues that are common to geo-participation practices. First, while geosocial media is a potential data source for understanding public opinions, it is unknown how much information related to public opinions or perceptions regarding general or specific policy and decision-making can be harvested and how this information might differ from public input collected from other participation approaches. In other words, if and how it can improve public participation or improve decision-making remains unclear. Second, there is a lack of empirical evidence on payoffs and uncertainties of using social media to enhance citizen participation and collaboration (Criado et al., 2013). Further studies are necessary to understand the practical challenges of government adopting geosocial media for participatory purposes.

1.3 Overview of the Book

Chapters 2and3 of this book synthesize recent advancements in spatially enabled ICT-driven citizen participation and propose a new typology for classifying and characterizing concepts and practices related to geospatial technology-mediated public participation. Each type of geo-participation practice is examined with practical examples to illustrate how new dynamics between governments and citizens are formed and new methods of collecting local spatial knowledge are enabled. The proposed typology is then applied to academic literature and government programs to demonstrate its use. The chapter concludes with emerging research and implementation needs.

Chapters 4and5 examine new opportunities for citizens to communicate transitory issues regarding the dynamic urban environment and to report non-emergency issues to local governments. A case study centered on sidewalk issues in the City of Kitchener (Ontario, Canada) compared citizen evaluations of the usefulness of geosocial media reporting with interview findings from staff of twelve Canadian municipalities. Bringing together government staff and citizen perspectives, the study presents key opportunities and challenges of using geosocial media reporting and suggests new forms of citizen collaboration to address data quality challenges and augment public service delivery.

Chapter 6 focuses on the potential for local governments or researchers to repurpose geosocial media data to gauge spatial and temporal dynamics of public opinions in ways that complement information collected through traditional public engagement methods. Text-processing methods are used to analyze Twitter data collected for the Region of Waterloo (Ontario, Canada) between March 2014 and July 2015 to assess citizens' concerns related to the planning and construction of a new Light Rail Transit (LRT) line. The case study illustrates how geosocial media can help identify geographies of expressed public concerns. The toolkit that is developed also addresses technical challenges of using the geosocial data that are usually unstructured, vary in quality, and require considerable effort to extract information that is relevant to local governments' needs.

Chapter 7 revisits recurring outcomes identified in the empirical studies, identifies the key contributions of the book and discusses directions for future academic research.

References

Anthopoulos, L., Reddick, C. G., Giannakidou, I., & Mavridis, N. (2014). Why e-government projects fail? An analysis of the Healthcare.gov website. *Government Information Quarterly, 33*(1), 161–173. https://doi.org/10.1016/j.giq.2015.07.003

Brandeis, M. W., & Nyerges, T. (2016). Assessing resistance to volunteered geographic information reporting within local government. *Transactions in GIS, 20*(2), 203–220. https://doi.org/10.1111/tgis.12168

Brown, G., & Kyttä, M. (2014). Key issues and research priorities for public participation GIS (PPGIS): A synthesis based on empirical research. *Applied Geography, 46,* 126–136. https://doi.org/10.1016/j.apgeog.2013.11.004

Cao, J., Zeng, K., Wang, H., & Cheng, J. (2014). Web-based traffic sentiment analysis: methods and applications. In *IEEE Transactions on Intelligent Transportation Systems* (Vol. 15, pp. 844–853).

Criado, J. I., Sandoval-Almazan, R., & Gil-Garcia, J. R. (2013). Government innovation through social media. *Government Information Quarterly, 30*(4), 319–326. https://doi.org/10.1016/j.giq.2013.10.003

De Longueville, B., Annoni, A., Schade, S., Ostlaender, N., & Whitmore, C. (2010). Digital Earth's nervous system for crisis events: Real-time sensor Web Enablement of volunteered geographic information. *International Journal of Digital Earth, 3*(3), 242–259. https://doi.org/10.1080/17538947.2010.484869

Johnson, P. A., & Sieber, R. E. (2012). Motivations driving government adoption of the Geoweb. *GeoJournal, 77*(5), 667–680. https://doi.org/10.1007/s10708-011-9416-8

Johnson, P. A., & Sieber, R. E. (2013). Situating the adoption of VGI by government. In D. Sui, S. Elwood, & M. Goodchild (Eds.), *Crowdsourcing geographic knowledge: Volun-teered geographic information (VGI) in theory and practice* (pp. 65–81). Springer. https://doi.org/10.1007/978-94-007-4587-2

Kaplan, A. M., & Haenlein, M. (2010). Users of the world, unite! The challenges and opportunities of Social Media. *Business Horizons, 53*(1), 59–68. https://doi.org/10.1016/j.bushor.2009.09.003

Khan, G. F., Yoon, H. Y., & Park, H. W. (2014). Social media communication strategies of government agencies: Twitter use in Korea and the USA. *Asian Journal of Communication, 24*(1), 60–78. https://doi.org/10.1080/01292986.2013.851723

Layne, K., & Lee, J. (2001). Developing fully functional E-government: A four stage model. *Government Information Quarterly, 18*(2), 122–136. https://doi.org/10.1016/S0740-624X(01)00066-1

Lev-On, A., & Steinfeld, N. (2015). Local engagement online: Municipal Facebook pages as hubs of interaction. *Government Information Quarterly, 32*(3), 299–307. https://doi.org/10.1016/j.giq.2015.05.007

Linders, Dennis. From e-government to we-government: Defining a typology for citizen coproduction in the age of social media. *Government information quarterly 29*.4 (2012): 446–454.

López-Ornelas, E., & Zaragoza, N. M. (2014). Social media participation: A narrative way to help planners. In *Social computing and social media* (pp. 48–54). https://doi.org/10.1007/978-3-319-07632-4

Magro, M. J. (2012). A Review of Social Media Use in E-Government. *Administrative Sciences, 2*(2), 148–161. https://doi.org/10.3390/admsci2020148

McKinsey Center for Government. (2017). *Government Productivity: Unlocking the $3.5 trillion opportunity.*

Mitchell, L., Frank, M. R., Harris, K. D., Dodds, P. S., & Danforth, C. M. (2013). The geography of happiness: connecting Twitter sentiment and expression, demographics, and objective characteristics of place. *PLoS ONE, 8*(5), Article e64417. https://doi.org/10.1371/journal.pone.0064417

Poorthuis, A., & Zook, M. (2017). Making big data small: strategies to expand Urban and geographical research using social media. *Journal of Urban Technology,* 1–21. Taylor & Francis. https://doi.org/10.1080/10630732.2017.1335153

Reddick, C. G., & Norris, D. F. (2013). Social media adoption at the American grass roots: Web 2.0 or 1.5? *Government Information Quarterly, 30*(4), 498–507. https://doi.org/10.1016/j.giq.2013.05.011

Rinner, C. (2006). Argumentation mapping in collaborative spatial decision making. In S. Balram & S. Dragićević (Eds.), *Collaborative GIS* (pp. 1–14).

Roche, Stéphane. Geographic Information Science I: Why does a smart city need to be spatially enabled?. *Progress in Human Geography 38*.5 (2014): 703–711.

Sawicki, D. S., & Peterman, D. R. (1996). Surveying the extent of PPGIS practice in the United States.

Sieber, R. E., Robinson, P. J., Johnson, P. A., & Corbett, J. M. (2016). Doing public participation on the Geospatial Web. *Annals of the American Association of Geographers, 106*(5), 1030–1046. https://doi.org/10.1080/24694452.2016.1191325

Sivarajah, U., Weerakkody, V., Waller, P., Lee, H., Irani, Z., Choi, Y., Morgan, R., & Glikman, Y. (2015). The role of e-participation and open data in evidence-based policy decision making in local government. *Journal of Organizational Computing and Electronic Commerce, 26*(1–2), 64–79. https://doi.org/10.1080/10919392.2015.1125171

Spinsanti, L., & Ostermann, F. (2013). Automated geographic context analysis for volunteered information. *Applied Geography, 43,* 36–44. https://doi.org/10.1016/j.apgeog.2013.05.005

Tang, Z., & Liu, T. (2015). Evaluating Internet-based public participation GIS (PPGIS) and volunteered geographic information (VGI) in environmental planning and management. *Journal of Environmental Planning and Management,* (August 2015), 1–18. https://doi.org/10.1080/09640568.2015.1054477

United Nations. (2008). United Nations E-Government Survey 2008: From E-government to connected governance.

United Nations. (2016). United Nations E-Government Survey 2016: E-government in support of sustainable development.

Chapter 2
Evolving Practices of Citizen Participation via Geospatial Technology: A Roadmap to Geo-Participation

Abstract Advances in spatially enabled information and communication technologies (ICTs) have provided governments with the potential to enhance public participation and to collaborate with citizens. This chapter reviews the concept of public participation and the progress of geo-enabled practices related to public participation (termed here as geo-participation) in government decision-making. The chapter suggests that the potential for local governments to embark on emerging geo-participation practices needs to be critically assessed.

2.1 Public Participation in Government Decision-Making

2.1.1 Defining Public Participation

Public participation by itself is a nebulous concept that is variously defined by its purposes, methods, and processes. In social science studies, for example, citizen participation is considered to be a collection of methods for involving the citizenry in the process of making decisions that affect them (Abelson et al., 2003). Xie and Jaeger (2008, p. 3) considered the aim of citizen participation is to shape governmental policy, "either by influencing the selection of government personnel or by affecting their choices." The OECD (2001) emphasized the active partnership of citizens and government in public participation processes for "defining the process and content of policy-making." The International Association of Public Participation (IAP2) considers the form of public participation and suggests five types of participation activities that indicate different levels of citizen involvement. The five types include informing, consulting, involving, collaborating, and empowering (IAP2, n.d.). Specifically, informing refers to government informing the citizenry about public policies, issues, and decisions; consulting refers to government collecting public feedback on decisions or plans; involving suggests that local governments carefully consider public interests and concerns in the decision-making processes; collaborating suggests that citizens may participate in developing criteria

S. Zhang, *Geo-participation*, https://doi.org/10.1007/978-981-95-8228-0_2

for decision-making; and empowering implies that the public has the right to make the final decisions.

Public participation is also often used in relation to other terms and concepts such as citizen or public engagement. Rowe and Frewer (2005), for instance, used public participation and public engagement interchangeably to refer to "the practice of involving members of the public in the agenda-setting, decision-making, and policy-forming activities of organizations/ institutions responsible for policy development" (Rowe & Frewer, 2005, p. 253). Other conceptualizations consider public participation and engagement to be relevant but different practices. For example, Ross et al. (2016) consider public engagement to be the broad term that includes public participation and other efforts of establishing public relations and media relations. Yet, their definition conflicts with the definition of IAP2 in that they consider public consultation to be different from public participation, whereas IAP2 defines public consultation as one form of public participation.

This book adopts the working definition of UNPAN (n.d.), which considers public participation to be the involvement of members of the public in a wide range of agenda-setting, decision and policy-making activities of local governments. The goal of participation is "to orient government programs toward community needs, build public support, and encourage a sense of cohesiveness within neighborhoods" (UNPAN, n.d.). Participation is not limited to particular forms but may encompass a spectrum of activities such as government informing citizenry of public policies and decisions, public consultation, and citizen collaboration on service provision and data collection.

According to the definition of UNPAN (n.d.), public participation may happen both within institutional arrangements and as grassroots initiatives. That is, participatory projects may be developed, guided, and facilitated by local government officials as part of the pre-established decision-making processes (Seeger, 2008). Citizens may also initiate their own projects to discuss issues of their communal concerns and to give their voices to public affairs (Lin, 2013; Panagiotopoulos et al., 2014). In contrast, public or citizen engagement refers only to collective and active citizen involvement through top-down participation initiatives led by government officials.

2.1.2 Rationale for Public Participation

Despite the challenges of defining public participation, few would argue with the fundamental belief in the democratic value of public participation. The International Association of Public Participation suggests that "those who are affected by a decision (should) have a right to be involved in the decision-making process" (IAP2, n.d.). The belief is that, at least in theory, decisions that are collectively made by citizens would reflect aggregated preferences of members of the public and would lead to public decisions that are democratic and legitimate (Innes & Booher, 2004). Others suggest that public participation is a means for eliminating unequal power relationships and advancing fairness and justice, as people who hold little power can have a voice (Fagence, 1977; Fischer, 1993). Participation is associated with empowerment as observed by Gaye and Diallo (1997, p. 12): "the newly empowered local community, through democratic decision-making and problem solving, matures into a body capable of interacting collectively with the local authority and even with agencies from higher government."

In addition to the democratic ideal, a more practical rationale for public participation is that it contributes positively to governmental decision-making by facilitating "responsive administrative apparatus as well as policy that is more representative of citizens' desires and needs" (Day, 1997, p. 425). It was reported that successful deliberative democracy projects can "vastly improve social outcomes, as balanced input from citizen participants allows factions to compromise and find solutions to previously intractable problems" (Irvin & Stansbury, 2004, p. 57). For government administrators, public participation may yield better public acceptance of decisions, as the public is better informed of the problems and the rationale behind the decision-making (Irvin & Stansbury, 2004). This improved public acceptance will also increase the possibility of successful implementation of decisions (Thomas, 1995). More recently, it has been suggested that involving citizens in government solutions can help improve the efficiency of government operations and also enhance government transparency (Linders, 2012).

2.1.3 Conceptualizing Public Participation

Public participation may happen at different stages of decision-making, be motivated by different purposes, adopt different participation mechanisms, induce different relationships among participants and decision-makers, and involve members of the public that reflect different compositions (Schlossberg & Shuford, 2005). Conceptualizations of public participation vary accordingly. Table 2.1 lists a number of widely adopted conceptualizations of public participation that focus on public participation as a means to enhance citizen power and control over decision-making, as an administrative process, and in terms of information flow. For each conceptualization, a spectrum of public participation activities ranging from low-level engagement (bottom) to high-level engagement (top) is classified according to the orientation of the conceptualization.

Perhaps the most prominent model of citizen participation is the ladder model suggested by Arnstein (1969). This model frames participation based on citizen power and classifies participation activities into eight rungs. These rungs correspond to three broad categories, namely nonparticipatory, tokenism, and citizen power. Along the spectrum, citizen powers range from no power to full control of decision-making. At the bottom level of the ladder, citizens do not have the power in decision-making; rather, those who have powers make the decision and elicit public support by educating the public. At the other end of the ladder, citizens have the power to influence the entire processes of decision-making and may even possess the ability to directly impact and control public decisions.

Table 2.1 Different conceptualizations of public participation

Orientation	Spectrum of public participation
Citizen power (Arnstein, 1969)	• Citizen power • Tokenism • Nonparticipation
Administrative (Wiedemann & Femers, 1993)	• Public partnership in the final decision • Public participation in assessing the risks and recommending solutions • Public participation in defining the interest and determining agenda • Public right to object • Inform the public • Public right to know
Information flow (Rowe & Frewer, 2005)	• Public participation • Public consultation • Public communication
E-participation (United Nations, 2008)	• E-participation • E-consultation • E-information

Wiedemann and Femers (1993) suggest an alternative ladder model that focuses on how public participation is programmed within government agencies. Their model accounts for how government agencies differently consider public participation in risk-related decision-making. For example, one may consider "distributing information to concerned citizens" as being public participation (Wiedemann & Femers, 1993, p. 356). In another case, public participation was considered to be the involvement of citizenry in assessing the risks of making a decision by public agencies. Accordingly, citizens are differently involved with government decision-making, ranging from "general education with little direct influence on decision-making to public participation in the final decision-making processes" (Schlossberg & Shuford, 2005, p. 17).

Rowe and Frewer (2005) frame public participation based on "the flow of information between participants and sponsors" (p. 254) and differentiate between three forms—public communication, public consultation, and public participation—that all are used for public participation. For public communication, information is conveyed from conveners to the public. For public consultation, information is conveyed from the public to conveners, and conveners should elicit and include public opinions in decision-making. Public participation suggests interactive information exchange between the public and the conveners. The dialogues between citizen representatives and conveners are believed to, to some degree, affect opinions of both parties and decision outcomes. Along the same line, The United Nations (2008) classifies e-participation activities into three types: e-informing (i.e., the one-way information provision from governments to citizens), e-consultation (i.e., the two-way communication through which governments solicit citizen opinions, input, and feedback), and e-decision-making (i.e., an interactive government–citizen relationship through which citizen participation may impact the decision-making process).

The public participation frameworks outlined here are just a few of the many conceptualizations of public participation. For example, others suggest participation models in the context of conflict resolution, preventing public controversy, and open government (Connor, 1988; Dorcey, 1994; Lee & Kwak, 2012). This list of public participation frameworks demonstrates the vast differences in orientations of participation and accordingly the acts of participation. As suggested by Schlossberg and Shuford (2005), the orientation or objectives of participation will determine how public participation is conceived, implemented, and evaluated.

2.1.4 Public Participation in Practice

In practice, the acts of participation vary according to the purpose of and the rationale for participation and have evolved with technological developments. Initial citizen participation projects in contemporary society were limited to small groups of citizen leaders (Day, 1997). The 1954 Urban Renewal Act in the United States, for example, established an advisory board that was comprised of seven to fifteen citizen leaders who had connections with people involved with urban renewal projects (e.g., contractors and bankers) as representatives of the public (Day, 1997). Starting in the 1960s, there was a transition from engaging a small elite group to engaging as many local communities and individuals that may be affected by a decision as possible (Callies, 1981). It was believed that ordinary citizens should have the right to give their voices in decisions that might impact them (Burke, 1979). The rise of ICTs has provided further opportunities to broaden public participation processes. Particularly, it was suggested that ICTs could help improve public access to government information and services and enable interactive dialogues among governments and citizens (Ho, 2002; Remmen, 2004). Accordingly, a wide range of digitally enabled participation activities including e-voting, online political discourse, online decision-making, e-activism, e-consultation, e-campaigning, and e-petitioning have emerged (Medaglia, 2012).

In more recent years, developments in Web 2.0 technologies provide increasing networking opportunities for governments to enhance public participation and enable wider-scale and new forms of government–citizen collaboration (Brabham, 2009). Web 2.0 technologies suggest a new paradigm of developing web technologies that highlight user-generated content, multidirectional information flows, and lightweight web application development cycles (O'Reilly, 2005). The roles of web users have shifted as they become both consumers and publishers of web content. According to Kaplan and Haenlein (2010), this paradigm shift has opened up the potential to achieve mass collaboration in that large numbers of web users can now collectively create content. In the government context, governments are taking advantage of the opportunity to enhance their participatory culture and embark on government–citizen collaboration initiatives to improve and augment public data and services (Brabham, 2009; Collins et al., 2013). The City of Chicago, for example, initiated a project called Snowportal for citizens to track snow-related information as well as to collectively update road and snow-removal conditions (Linders, 2012). The collective capabilities of the public to address the challenge of real-time data collection offer the potential for governments to reduce the financial, time, and human costs of providing public services and to improve the efficiency of service provision.

2.1.5 Challenges of Implementing Public Participation

Despite the continuing efforts to enhance public participation, implementing public participation still faces challenges. The concern that the outcomes of a participation process will reflect the preferences or interests of only certain population groups has been longstanding (Day, 1997). For citizens, participation requires resources such as time, money, and energy; people who possess more resources are more likely to participate than those who possess fewer (McCarthy & Zald, 2001, 1977). Concerns of excluding socially disadvantaged people have been widely discussed as these people may not have the resources necessary to participate (Cinderby, 2010; Sieber, 2006). Some of the barriers to participation, such as the confines of space and time, are mitigated by Internet technologies, as people can participate remotely at times of their choice (Vicente & Novo, 2014). However, new technical and social barriers have emerged, as people need access to digital devices (e.g., computers) and must possess a certain level of digital and technical skills to participate (Crutcher & Zook, 2009; Lin, 2013). Others have observed generational gaps in that younger generations are found to be more likely to accept digital participation methods (Corbett, 2013; Tulloch, 2008). The concern of excluding socially disadvantaged groups remains as these people may still lack necessary resources to digitally participate (Vicente & Novo, 2014).

In addition to the concern of a biased representation of the public, Etzioni-Halevy (2013) warns that a decentralized public participation process does not always equally empower citizens to share their concerns and aspirations. Smith and McDonough (2001) provided evidence that citizens are often not satisfied with participation processes, as citizens noticed that participation was unfair and participation outcomes were directed to particular results favored by elite groups or governments. In online environments, studies have similarly suggested that the free and self-expression channel does not eliminate inequality but is biased in favor of opinion leaders who have significant influences on online communications (Hong, 2013). Consequently, citizens may obtain a sense of social exclusion or have negative views of local authorities and would be less likely to participate (Lowndes et al., 2001; Roeder et al., 2005).

Moreover, public participation outcomes do not always impact decision-making processes. According to an e-government survey conducted by the United Nations, "only 38 countries out of 193 member states (20%) indicate that e-consultation outcomes have resulted in new policy decisions, regulations, or service" (United Nations, 2016, p. 65). The institutional settings of local government, including its political structures, routines, and cultures, are found to be counterproductive to public participation (Colombo, 2010; Grönlund, 2009). Operations in a bureaucratic organization such as government need to follow "a set of formal, explicit, comprehensive and stable rules that are impersonally enforced in decision-making and lead to predictable and determinate results" (Cordella & Tempini, 2015, p. 280). With inadequate representation of citizen interest, governments face risks of producing decisions that are unduly influenced by citizenry representatives (Irvin & Stansbury, 2004). As a result, public agencies are found to be more likely to trust their own capabilities of influencing decision-making from internal processes (Brown, 2012).

As such, the implementation of public participation remains challenging. It is suggested that further research and practices should pay attention to improving public engagement processes (Brown & Kyttä, 2014; Haworth et al., 2016) and seek technological solutions to broaden the participation process and encourage public collaboration for tackling urban problems (Brabham, 2009; Bright & Margetts, 2016).

2.2 Geo-Participation

The incorporation of geospatial information and methods in public participation was motivated by the goals of: (1) educating the public on the spatial context of decisions, (2) bringing in local, experiential knowledge that often has a spatial component such as important locations for a planning or development proposal to decision-making processes, and (3) understanding geographies of public perceptions toward decisions made by public agencies and authorities (Brown & Kyttä, 2014; Kingston, 2007; Sieber, 2006).

2.2.1 Public Participation GIS

In the 1990s, the term public participation GIS (PPGIS) was proposed to describe uses of GIS systems as tools for government agencies to enhance public involvement (Brown & Kyttä, 2014; Kar et al., 2016). Numerous studies have reported applications of PPGIS in a wide range of participatory planning projects for land uses, forest, marine, and community development (Atzmanstorfer et al., 2014; Brown, 2009; Brown & Brabyn, 2012; St. Martin & Hall-Arber, 2008). PPGIS was articulated as a well-established technical method for citizens to use to "map alternate views of the same problem and analyze the same data differently from those with political power" (Kar et al., 2016, p. 296) and for the public, stakeholders, and decision-makers to use to communicate and even collaborate through creating and disseminating map-based information (Hall et al., 2010).

PPGIS is not merely tool development. An important thread throughout PPGIS studies is whether or not the technology would empower local communities and individuals, include marginalized groups in decision-making processes, and increase public acceptability of authoritative decisions (Brown, 2017; Corbett et al., 2016). Studies have suggested steady progress in improving public access to map-based information (Sawicki & Peterman, 1996), engaging with "hard-to-reach" groups (Cinderby, 2010) and the senior population (Gottwald et al., 2016), and increasing social acceptability of land use decisions (Brown, 2017). Yet concerns remain that the representation of the public in PPGIS approaches is biased toward those comfortable using map-based interfaces and also that the technologically driven approaches benefit some communities and individuals more than others (Brown et al., 2014; Elwood & Ghose, 2011).

2.2.2 The Geoweb and Public Participation

Influenced by Web 2.0 technologies, the Geoweb, a collection of online geospatial technologies that support online generation, dissemination, and management of geospatial data, information, and maps, emerged (Elwood, 2010). The Geoweb brought a significant paradigmatic shift in spatial data production (Elwood & Leszczynski, 2013). With increasing use of location-aware devices (e.g., smartphones, wearable devices, and GPS-enabled sensors), ubiquitous access to the Internet, and growing popularity of web mapping applications (e.g., OpenStreetMap) and location-based services, generating spatial data is no longer exclusive to trained experts but has become accessible to average citizens (Goodchild, 2007). On the one hand, the paradigm shift has greatly expanded the capabilities of PPGIS systems, as people now have fewer physical (e.g., space and time) and technical barriers to using online mapping applications (Tang & Liu, 2015). On the other hand, the paradigm shift has given rise to new forms of participation that differ significantly from PPGIS in terms of their purpose, processes, and methods, and provides local governments with the potential to enhance public participation and spur citizen collaboration and innovation (Sieber et al., 2016; Unsworth et al., 2014).

The Geoweb technologies have reduced technical barriers for laypeople to create maps and have encouraged participation with little or no formal coordination (Sieber et al., 2016). Lin (2013), for example, suggests that practices such as neogeographic mapping enable local communities or individuals to initiate participatory activities and express their opinions through collective mapping outside the traditional arena of participation. Crowdsourcing mapping applications, such as OpenStreetMap, are successful examples of community-initiated collective efforts to generate and validate geospatial information that can be used by individuals, communities, and private and public sectors (Neis & Zielstra, 2014).

The Geoweb has spurred new forms of government–citizen collaboration. Mirroring the vision of Goodchild (2007), that citizens could act as sensors to revitalize costly expert-driven data systems and to collectively observe a phenomenon and possibly respond faster to the phenomenon, a full spectrum of knowledge producers (e.g., from expert cartographers to lay public) may collectively contribute to government data, services, and infrastructure. For example, citizens may provide up-to-date information (e.g., snow, flooding, traffic) and collaborate with governments on providing customized geolocation services (Linders, 2012). Inspired by successful crowdsourcing mapping applications, there is potential for governments to embark on collective mapping practices and involve citizens in updating, correcting, and vetting government geospatial datasets (Johnson, 2016). According to Sieber and Johnson (2015), this collaboration can be considered a participatory model where data becomes a conduit for integrating citizen contribution into government operations and decision-making. Citizens could also collaborate with governments to produce public services and applications for solving urban problems. According to Desouza and Bhagwatwar (2012), citizens collaborate with governments for problem identification and resolution through citizen apps on a variety of topics including transportation, public utilities, health and recreation, and information awareness and access.

A new form of passive citizen sensing has also emerged because of abundant geospatial data generated through the Geoweb. That is, citizens generate geospatial information when using GPS-enabled devices and online location-based services without a participatory intention. Yet, the information can be used to obtain aggregated local spatial knowledge related to lived experiences, public perceptions, and local environments (Crooks et al., 2013). Volunteered geographic information (VGI), geospatial content contributed to the Geoweb by web users, for example, is increasingly adopted to obtain information about time-critical situations and help enhance situation awareness (Hughes & Palen, 2009; Yin et al., 2012). In the local government context, passive citizen sensing has the potential to supplement or augment existing public participation approaches. Gao et al. (2017), for instance, found that public perceptions extracted from social media are comparable to those collected using in-person techniques.

Local governments may adopt these emerging practices and opportunities for different purposes and uses. For example, governments may feed citizen-generated data to "higher-level decision-makers in a one-way process" or interactively engage with citizens in a two-way dialogue to improve their efficiency, transparency, and accountability (Johnson & Sieber, 2013, p. 77). Applications of some practices are still in their early stages. Citizen collaboration on vetting government open data (considered as participatory open data by Sieber and Johnson, 2015), for example, is an initiative that governments are only beginning to embark on. Therefore, further studies are necessary to assess the participation potential of these practices, investigate how these practices may fit into government procedures, and determine the best practices for government decision-making.

2.2.3 Defining Geo-Participation

Various terms are used for the emerging practices spurred by the Geoweb, including VGI (Goodchild, 2007), citizen-generated spatial data (Mooney et al., 2011), neogeographic mapping (Turner, 2006), crowdsourced mapping (Miller & Goodchild, 2015), civic issue trackers (Sieber & Johnson, 2015), citizen sensing (Schade et al., 2013), and social sensing (Liu et al., 2015), to name a few. In the context of participation, this multiplicity of terms raises confusion as the terms are often used interchangeably, the lines between concepts and practices are often blurred, and the participatory contexts and objectives they involve are often not clearly defined. Geo-participation is used in this book to refer to a collection of practices that are "infused with geospatial and participatory potential" (Kar et al., 2016, p. 296).

Some may argue that these practices are not wholly developed based on the intention of participation nor compatible with traditional conceptualizations of public participation. For example, VGI primarily focuses on generating and disseminating geographic information, which may or may not have a participation goal (Verplanke et al., 2016). Some types of citizen sensing (e.g., passive citizen sensing, ambient VGI) may not involve active public engagement that is usually key to public participation processes but relates to mining aggregated public perceptions using data-driven approaches (Gao et al., 2017). However, these practices can be used for participatory purposes, as governments may benefit from these practices in terms of broadened public engagement, better understanding of public needs and improvements of public services and decision-making (Bright & Margetts 2016).

2.2.4 Geo-Participation: Challenges and Research Issues

Geo-participation encompasses a range of practices that adopt different tools, involve different participation processes, and may produce different participation outcomes. For example, some VGI practices facilitate direct map-based discussions between citizens and decision-makers and should be considered along the lines of traditional PPGIS research (Tang & Liu, 2015). Other practices deviate from traditional conceptualizations of public participation, as they may have an orientation (e.g., enriching government data) that is not emphasized in traditional conceptualizations of public participation based on citizen powers, conflict resolutions, and administrative mandates. As a result, new forms of government–citizen interaction may emerge that cannot be conceptualized by the traditional ladder metaphor (Sieber et al., 2016). To assess the participation potential of geo-participation, it is necessary to classify the emerging geo-participation practices and identify the participation purposes, processes, and outcomes associated with each type of geo-participation.

Based on the classification, we should then identify and assess if and how geo-participation can enhance citizen participation and improve decision-making. First, while studies suggest that the Geoweb may increase the number of participants and bring in a more diverse composition of participants (Brown, Kelly, et al., 2014; Lin, 2013), Elwood and Leszczynski (2013) warn us that digital divides are likely to persist with technological developments, as people may use technology differently due to nonuniform motivation, access, and skills. Alternatives such as mining public opinions using online sources may help us to solicit citizen input from some underrepresented groups (e.g., youth) and obtain unfiltered public views (Dunkel, 2015; Schweitzer, 2014). However, this potential needs to be further assessed in terms of its usefulness (e.g., what information can be elicited) and in terms of government acceptance of such an approach (Bright & Margetts, 2016).

Section 2.1.5 suggests that one challenge of public participation is that participation outcomes may not be incorporated into decision-making. According to Brown and Kyttä (2014), uncertain data quality of contributed data is one factor that may impede the integration of citizen contribution into official decision-making. With wider-scale citizen contribution through the Geoweb, varying geographic knowledge, diverging perceptions of on-the-ground situations, diverse technical skills, and less agreement on terminologies by contributors further exacerbate the consistency of heterogeneous datasets (Bakillah et al., 2013; Senaratne et al., 2017). This presents significant technical challenges for governments to integrate contributed data into their operations and decision-making, as governments need to develop mechanisms of validating data quality, managing heterogeneous datasets, and supporting interoperability among datasets (Bakillah et al., 2013; Garnett & Kanaroglou, 2016). Moreover, the anonymity of online participation raises an issue of trustworthiness for governments to adopt contributed data (Johnson, 2016). PPGIS approaches emphasize the localness of participants, as they would bring in local knowledge to decision-making. With online participation, the extent to which anonymous participants are familiar with local issues is unknown to local governments. Further studies are necessary to identify proper mechanisms that can assure quality of citizen input yet not deteriorate the intention of broadening participation (Brabham, 2009).

It is also necessary to examine how geo-participation practices may fit into local government procedures by addressing various organizational, institutional, technical and legislative factors. In practice, local governments need to consider how many human and financial resources need to be allocated for the new practice; whether the new practice conflicts with existing frameworks and procedures; if new strategies, procedures, policies, and legislation need to be established; and whether new tools and systems need to be developed and how they are compatible with the legacy IT systems of local governments (Bertot et al., 2016; Janowski, 2015; Johnson & Sieber, 2013; Nam & Pardo, 2014). As such, authorities' perspectives are critical to the examination and assessment of geo-participation practices.

References

Abelson, J., Forest, P. G., Eyles, J., Smith, P., Martin, E., & Gauvin, F. P. (2003). Deliberations about deliberative methods: Issues in the design and evaluation of public participation processes. *Social Science & Medicine, 57*(2), 239–251.

Arnstein, S. R. (1969). A ladder of Citizen Participation. *Journal of the American Institute of Planners, 35*(4), 216–224. https://doi.org/10.1080/01944366908977225

Atzmanstorfer, K., Resl, R., Eitzinger, A., & Izurieta, X. (2014). The GeoCitizen-approach: community-based spatial planning–an Ecuadorian case study. *Cartography and geographic information science, 41*(3), 248–259.

Bakillah, M., Liang, S., Zipf, A., & Arsanjani, J. (2013). Semantic Interoperability of Sensor Data with Volunteered Geographic Information: A Unified Model. *ISPRS International Journal of Geo-Information, 2*, 766–796. http://doi.org/10.3390/ijgi2030766

Bertot, J., Estevez, E., & Janowski, T. (2016). Universal and contextualized public services: Digital public service innovation framework. *Government information quarterly, 33*(2), 211–222.

Brabham, D. C. (2009). Crowdsourcing the public participation process for planning projects. *Planning Theory, 8*(3), 242–262. https://doi.org/10.1177/1473095209104824

Bright, J., & Margetts, H. (2016). Big Data and Public Policy : Can It Succeed Where E-Participation Has Failed ? *Policy and Internet, 8*(3), 218–224.

Brown, G., Kelly, M., & Whitall, D. (2014). Which 'public'? Sampling effects in public participation GIS (PPGIS) and volunteered geographic information (VGI) systems for public lands management. *Journal of Environmental Planning and Management, 57*(2), 190–214.

Brown, G., & Kyttä, M. (2014). Key issues and research priorities for public participation GIS (PPGIS): A synthesis based on empirical research. *Applied Geography, 46*, 126–136. http://doi.org/10.1016/j.apgeog.2013.11.004

Brown, G., & Brabyn, L. (2012). An analysis of the relationships between multiple values and physical landscapes at a regional scale using public participation GIS and landscape character classification. *Landscape and urban planning, 107*(3), 317–331.

Brown, G. (2009). Public participation GIS: A new method for use in national forest planning. *Forest Science, 55*(2), 166–182.

Brown, G. (2012). Public participation GIS (PPGIS) for regional and Environmental Planning: reflections on a decade of empirical research. *URISA Journal, 25*(2), 7–18.

Brown, G. (2017). A review of sampling effects and response bias in Internet participatory mapping (PPGIS/PGIs/VGI). *Transactions in GIS, 21*(1), 39–56. https://doi.org/10.1111/tgis.12207

Burke, E. M. (1979). *A participatory approach to urban planning*. Human Sciences Press.

Callies, D. (1981). Public Participation in the United States. *Town Planning Review, 52*(3), 286–296.

Cinderby, S. (2010). How to reach the "hard-to-reach": The development of Participatory Geographic Information Systems (P-GIS) for inclusive urban design in UK cities. *Area, 42*(2), 239–251. https://doi.org/10.1111/j.1475-4762.2009.00912.x

Collins, L., Swart, R., & Zhang, B. (2013). *The rise of future finance: The UK alternative finance benchmarking report*. Nesta; University of Cambridge.

Colombo, C. (2010). E-participation experiences and local government in Catalonia: An explanatory analysis. In *Lecture Notes in Computer Science (including subseries Lecture Notes in Artificial Intelligence and Lecture Notes in Bioinformatics)* (Vol. 6229 LNCS, pp. 82–94). https://doi.org/10.1007/978-3-642-15158-3_7

Connor, D. M. (1988). A new ladder of citizen participation. *National Civic Review, 77*(3), 249–257. https://doi.org/10.1002/ncr.4100770309

Corbett, J., Cochrane, L., & Gill, M. (2016). Powering up: Revisiting participatory GIS and empowerment. *The Cartographic Journal, 53*(4), 335–340.

Corbett, J. (2013). "I don't come from anywhere": Exploring the role of the Geoweb and volunteered geographic information in rediscovering a sense of place in a dispersed Aboriginal community. In D. Sui, S. Elwood, & M. Goodchild (Eds.), *Crowdsourcing geographic knowledge: Volunteered geographic information (VGI) in theory and practice* (pp. 223–241). Springer. https://doi.org/10.1007/978-94-007-4587-2

Cordella, A., & Tempini, N. (2015). E-government and organizational change: Reappraising the role of ICT and bureaucracy in public service delivery. *Government Information Quarterly, 32*(3), 279–286. http://doi.org/10.1016/j.giq.2015.03.005

Crooks, A., Croitoru, A., Stefanidis, A., & Radzikowski, J. (2013). # Earthquake: Twitter as a distributed sensor system. *Transactions in GIS, 17*(1), 124–147.

Crutcher, M., & Zook, M. (2009). Placemarks and waterlines: Racialized cyberscapes in post-Katrina Google Earth. *Geoforum, 40*(4), 523–534. https://doi.org/10.1016/j.geoforum.2009.01.003

Day, D. (1997). Citizen participation in the planning process: An essentially contested concept? *Journal of Planning Literature, 11*(3), 421–434. https://doi.org/10.1177/088541229701100309

Desouza, K. C., & Bhagwatwar, A. (2012). Citizen Apps to Solve Complex Urban Problems. *Journal of Urban Technology, 19*(3), 107–136.

Dorcey, A. H. J. (1994). *Public involvement in government decision making: Choosing the right model: A report of the B.C. Round Table on the environment and the economy*. Victoria, B.C.

Dunkel, A. (2015). Visualizing the perceived environment using crowdsourced photo geodata. *Landscape and Urban Planning, 142,* 173–186. https://doi.org/10.1016/j.landurbplan.2015.02.022

Elwood, S., & Ghose, R. (2011). PPGIS in Community Development Planning: Framing the Organizational Context. *In Classics in Cartography*, M. Dodge (Ed.). https://doi.org/10.1002/9780470669488.ch8

Elwood, S., & Leszczynski, A. (2013). New spatial media, new knowledge politics. *Transactions of the Institute of British Geographers, 38*(4), 544–559.

Elwood, S. (2010). Geographic information science: Emerging research on the societal implications of the geospatial web. *Progress in Human Geography, 34*(3), 349–357. https://doi.org/10.1177/0309132509340711

Etzioni-Halevy, E. (2013). *Bureaucracy and democracy: A political dilemma* (Vol. 7). Routledge. https://doi.org/10.4324/9780203707890

Fagence, M. (1977). Citizen participation in planning. *Citizen Participation in Planning.* https://doi.org/10.1016/B978-0-08-020398-0.50009-7

Fischer, F. (1993). Citizen participation and the democratization of policy expertise: From theoretical inquiry to practical cases. *Policy Sciences, 26*(3), 165–187. https://doi.org/10.1007/BF00999715

Gao, S., Janowicz, K., Montello, D. R., Hu, Y., Yang, J., McKenzie, G., … Yan, B. (2017). A data-synthesis-driven method for detecting and extracting vague cognitive regions. *International Journal of Geographical Information Science, 31*(6), 1245–1271. http://doi.org/10.1080/13658816.2016.1273357

Garnett, R., & Kanaroglou, P. (2016). Qualitative GIS: An Open Framework Using SpatiaLite and Open Source GIS. *Transactions in GIS, 20*(1), 144–159. https://doi.org/10.1111/tgis.12163

Gaye, M., & Diallo, F. (1997). Community participation in the management of the urban environment in Rufisque (Senegal). *Environment and Urbanization, 9*(1), 9–30.

Gottwald, S., Laatikainen, T. E., & Kyttä, M. (2016). Exploring the usability of PPGIS among older adults: challenges and opportunities. *International Journal of Geographical Information Science, 30*(12), 2321–2338.

Goodchild, M. F. (2007). Citizens as sensors: The world of volunteered geography. *GeoJournal, 69*(4), 211–221. https://doi.org/10.1007/s10708-007-9111-y

Grönlund, Å. (2009, September). ICT is not participation is not democracy–eParticipation development models revisited. In *International Conference on Electronic Participation* (pp. 12–23). Berlin, Heidelberg: Springer Berlin Heidelberg.

Hall, G. B., Chipeniuk, R., Feick, R. D., Leahy, M. G., & Deparday, V. (2010). Community-based production of geographic information using open source software and Web 2.0. *International journal of geographical information science, 24*(5), 761–781.

Haworth, Billy, Joshua Whittaker, and Eleanor Bruce. 2016. "Assessing the Application and Value of Participatory Mapping for Community Bushfire Preparation." *Applied Geography 76*:115–27. http://dx.doi.org/10.1016/j.apgeog.2016.09.019.

Ho, A. T. (2002). Reinventing Local Governments and the E-Government Initiative. *Public Administration Review, 62*(4), 434–444. https://doi.org/10.1111/0033-3352.00197

Hong, S. (2013). Who benefits from Twitter? Social media and political competition in the U.S. House of Representatives. *Government Information Quarterly, 30*(4), 464–472. https://doi.org/10.1016/j.giq.2013.05.009

Hughes, A. L., & Palen, L. (2009). Twitter adoption and use in mass convergence and emergency events. *International journal of emergency management*, 6(3–4), 248–260.

Innes, J. E., & Booher, D. E. (2004). Reframing public participation: strategies for the 21st century. *Planning theory & practice, 5*(4), 419–436.

Irvin, R. a., & Stansbury, J. (2004). Citizen participation in decision making: Is it worth the effort? *Public Administration Review, 64*(1), 55–65. http://doi.org/10.1111/j.1540-6210.2004.00346.x

Janowski, T. (2015). Digital government evolution: From transformation to contextualization. *Government Information Quarterly, 32*(3), 221–236. https://doi.org/10.1016/j.giq.2015.07.001

Johnson, P.A., Sieber, R.E. (2013). Situating the Adoption of VGI by Government. In: Sui, D., Elwood, S., Goodchild, M. (eds) *Crowdsourcing Geographic Knowledge*. Springer, Dordrecht. https://doi.org/10.1007/978-94-007-4587-2_5

Johnson, P. A. (2016). Models of direct editing of government spatial data: Challenges and constraints to the acceptance of contributed data. *Cartography and Geographic Information Science, 406*(May), 1–11. https://doi.org/10.1080/15230406.2016.1176536

Kaplan, A. M., & Haenlein, M. (2010). Users of the world, unite! The challenges and opportunities of Social Media. *Business Horizons, 53*(1), 59–68. http://doi.org/10.1016/j.bushor.2009.09.003

Kar, B., Sieber, R., Haklay, M., & Ghose, R. (2016). Public participation GIS and participatory GIS in the era of GeoWeb. *The Cartographic Journal, 53*(4), 296–299.

Kingston, R. (2007). Public participation in local policy decision-making: The role of web-based mapping. *The Cartographic Journal, 44*(2), 138–144. https://doi.org/10.1179/000870407X213459

Lee, G., & Kwak, Y. H. (2012). An Open Government Maturity Model for social media-based public engagement. *Government Information Quarterly, 29*(4), 492–503. http://doi.org/10.1016/j.giq.2012.06.001

Liu, Y., Liu, X., Gao, S., Gong, L., Kang, C., Zhi, Y., … Shi, L. (2015). Social Sensing: A New Approach to Understanding Our Socioeconomic Environments. *Annals of the Association of American Geographers, 105*(3), 512–530. http://doi.org/10.1080/00045608.2015.1018773

Lin, W. (2013). Situating performative neogeography: Tracing, mapping, and performing "Everyone's East Lake." *Environment and Planning A, 45*(1), 37–54. https://doi.org/10.1068/a45161

Linders, D. (2012). From e-government to we-government: Defining a typology for citizen coproduction in the age of social media. *Government Information Quarterly, 29*(4), 446–454. https://doi.org/10.1016/j.giq.2012.06.003

Lowndes, V., Pratchett, L., & Stoker, G. (2001). Trends in public participation: part 1-local government perspectives. *Public administration, 79*(1), 205–222.

McCarthy, J. D., & Zald, M. N. (1977). Resource mobilization and social movements: A partial theory. *American journal of sociology, 82*(6), 1212–1241.

McCarthy, J. D., & Zald, M. N. (2001). The enduring vitality of the resource mobilization theory of social movements. In *Handbook of sociological theory* (pp. 533–565). Boston, MA: Springer US.

Medaglia, R. (2012). E-participation research: Moving characterization forward (2006–2011). *Government Information Quarterly, 29*(3), 346–360. https://doi.org/10.1016/j.giq.2012.02.010

Miller, H. J., & Goodchild, M. F. (2015). Data-driven geography. *GeoJournal*, (October 2014), 449–461. http://doi.org/10.1007/s10708-014-9602-6

Mooney, P., Sun, H., Corcoran, P., & Yan, L. (2011). Citizen-generated spatial data and information: Risks and opportunities. *Proceedings of 2011 IEEE International Conference on Intelligence and Security Informatics*, 232–232. http://doi.org/10.1109/ISI.2011.5984087

Nam, T., & Pardo, T. A. (2014). The changing face of a city government: A case study of Philly311. *Government Information Quarterly, 31*, S1–S9.

Neis, P., & Zielstra, D. (2014). Recent Developments and Future Trends in Volunteered Geographic Information Research: The Case of OpenStreetMap. *Future Internet, 6*(1), 76–106. http://doi.org/10.3390/fi6010076

OECD. (2001). *Citizen as partners: Information, consultation and public participation in policy-making*. OECD Publishing

O'Reilly, T. (2005). O'Reilly: *What Is Web 2.0*. Retrieved from http://www.oreillynet.com/pub/a/oreilly/tim/news/2005/09/30/what-is-web-20.html

Panagiotopoulos, P., Bigdeli, A. Z., & Sams, S. (2014). Citizen–government collaboration on social media: The case of Twitter in the 2011 riots in England. *Government information quarterly, 31*(3), 349–357.

Remmen, A. (2004). *Images of E-Government—Experiences from the digital North Denmark*. Aalborg University.

Roeder, S., Poppenborg, A., Michaelis, S., Märker, O., & Salz, S. R. (2005, March). "Public budget dialogue"–an innovative approach to e-participation. In *International Conference on e-Government* (pp. 48–56). Berlin, Heidelberg: Springer Berlin Heidelberg.

Ross, H., Baldwin, C., & Carter, R. W. (2016). Subtle implications: public participation versus community engagement in environmental decision-making. *Australasian Journal of Environmental Management, 23*(2), 123–129.

Rowe, G., & Frewer, L. J. (2005). A typology of public engagement mechanisms. *Science, Technology & Human Values, 30*(2), 251–290. http://doi.org/10.1177/0162243904271724

Sawicki, D. S., & Peterman, D. R. (1996). Surveying the extent of PPGIS practice in the United States.

Schade, S., Díaz, L., Ostermann, F., Spinsanti, L., Luraschi, G., Cox, S., … Longueville, B. (2013). Citizen-based sensing of crisis events: sensor web enablement for volunteered geographic information. *Applied Geomatics, 5*(1), 3–18. http://doi.org/10.1007/s12518-011-0056-y

Schlossberg, M., & Shuford, E. (2005). Delineating "Public" and "Participation" in PPGIS. *URISA Journal, 16*(2), 15–26. Retrieved from https://scholarsbank.uoregon.edu/xmlui/handle/1794/1343

Schweitzer, L. (2014). Planning and social media: A case study of public transit and stigma on Twitter. *Journal of the American Planning Association, 80*(3), 218–238. https://doi.org/10.1080/01944363.2014.980439

Senaratne, H., Mobasheri, A., Ali, A. L., Capineri, C., & Haklay, M. (2017). A review of volunteered geographic information quality assessment methods. *International Journal of Geographical Information Science, 31*(1), 139–167.

Seeger, C. J. (2008). The role of facilitated volunteered geographic information in the landscape planning and site design process. *GeoJournal, 72*(3–4), 199–213. https://doi.org/10.1007/s10708-008-9184-2

Sieber, R. E., Robinson, P. J., Johnson, P. A., & Corbett, J. M. (2016). Doing Public Participation on the Geospatial Web. *Annals of the American Association of Geographers, 106*(5), 1030–1046. http://doi.org/10.1080/24694452.2016.1191325

Sieber, R. E., & Johnson, P. A. (2015). Civic open data at a crossroads: Dominant models and current challenges. *Government Information Quarterly, 32*(3), 308–315. http://doi.org/10.1016/j.giq.2015.05.003

Sieber, R. E. (2006). Public participation geographic information systems: A literature review and framework. *Annals of the Association of American Geographers, 96*(3), 491–507. https://doi.org/10.1111/j.1467-8306.2006.00702.x

Smith, P. D., & McDonough, M. H. (2001). Beyond public participation: Fairness in natural resource decision making. *Society & natural resources, 14*(3), 239–249.

St. Martin, K., & Hall-Arber, M. (2008). The missing layer: Geo-technologies, communities, and implications for marine spatial planning. *Marine Policy, 32*(5), 779–786. http://doi.org/10.1016/j.marpol.2008.03.015

Tang, Z., & Liu, T. (2015). Evaluating Internet-based public participation GIS (PPGIS) and volunteered geographic information (VGI) in environmental planning and management. *Journal of Environmental Planning and Management*, (August 2015), 1–18. http://doi.org/10.1080/09640568.2015.1054477

Thomas, J. C. (1995). *Public participation in public decisions*. Jossey-Bass.

Tulloch, D. L. (2008). Is VGI participation? From vernal pools to video games. *GeoJournal, 72*(3–4), 161–171. https://doi.org/10.1007/s10708-008-9185-1

Turner, A. J. (2006). Introduction to Neogeography. *OReilly Short Cuts*. Retrieved from http://books.google.com/books?hl=en&lr=&id=oHgDv4feV-8C&oi=fnd&pg=PA24&dq=Introduction+to+Neogeography&ots=wYr7RESSeW&sig=HTani7IqwF0NQmiqf8jZigkFguQ

United Nations. (2008). *United Nations E-Government Survey 2008: From E-Government to Connected Governance*. New York.

United Nations. (2016). *United Nations E-government survey 2016: E-Government in support of sustainable development*.

Unsworth, K., Forte, A., & Dilworth, R. (2014). Urban informatics: The role of citizen participation in policy making. *Journal of Urban Technology, 21*(4), 1–5.

Verplanke, J., McCall, M. K., Uberhuaga, C., Rambaldi, G., & Haklay, M. (2016). A shared perspective for PGIS and VGI. *The Cartographic Journal, 53*(4), 308–317.

Vicente, M. R., & Novo, A. (2014). An empirical analysis of e-participation. The role of social networks and e-government over citizens' online engagement. *Government Information Quarterly, 31*(3), 379–387. https://doi.org/10.1016/j.giq.2013.12.006.

Wiedemann, P. M., & Femersb, S. (1993). Public participation in waste management decision making : Analysis and management of conflicts*Journal Hazardous Materials, 33*, 355–368

Xie, B., & Jaeger, P. T. (2008). Older adults and political participation on the internet: A cross-cultural comparison of the USA and China. *Journal of cross-cultural gerontology, 23*(1), 1–15.

Yin, J., Lampert, A., Cameron, M., Robinson, B., & Power, R. (2012). Using social media to enhance emergency situation awareness. *IEEE intelligent systems, 27*(6), 52–59.

Chapter 3
Defining a New Typology for Geo-Participation

Abstract Advancements in citizen sensing and geospatial big data have enabled new opportunities for government–citizen interactions and have played important roles in developing smart(er) cities. In addition to city governments and citizens actively using maps to communicate spatial planning issues, the increasing capabilities of citizens generating spatial data either actively or passively allow city governments to collect local spatial knowledge with unprecedented breadth at finer spatiotemporal resolutions. New methods for citizens collaborating with city governments are also emerging to enhance citizen engagement and spur social innovation. By synthesizing recent advancements in geo-enabled citizen participation, this chapter proposes a new typology for classifying and characterizing concepts and practices related to geospatial technology-mediated public participation in local governments (e.g., city and municipal governments). Practical examples are used to illustrate how new dynamics between local governments and citizens are formed, new methods of collecting local spatial knowledge are enabled, and new opportunities for improving the openness and operational efficiency of local governments have emerged. The proposed conceptualization and examples give rise to emerging needs of advancing geo-participation by developing geospatial methods and infrastructure and by investigating the social and spatial implications of geo-participation.

3.1 Introduction

With the growing digital culture in our society, content generation no longer takes place only within the wall of mainstream organizations (O'Reilly, 2005). The more networked digital media enables new forms of mass collaboration among individuals and private and public sectors (Brabham, 2009). Against this backdrop, governments are changing how they operate. According to The United Nations, there is a global trend of governments seeking for Internet-based participatory solutions to improve their service delivery, enhance citizen participation, and increase their transparency, openness and accountability (United Nations, 2016).

S. Zhang, *Geo-participation*, https://doi.org/10.1007/978-981-95-8228-0_3

Developments in geospatial technologies have contributed to these government efforts. Using map-based approaches to engage with the public and to seek participatory solutions for spatial planning decisions by incorporating local spatial knowledge is not new. In the 1990s, the term public participation GIS (PPGIS) was proposed to describe uses of GIS systems as tools for government agencies to enhance public involvement (Carver et al., 2001; Sieber, 2006). With increased access to Internet and location-aware technologies, technical barriers of participation have been reduced and a wider scale spatial data generation by ordinary citizens has been enabled. Goodchild (2007) proposes the term volunteered geographic information (VGI) to describe transforming roles of lay people to become spatial data creators. The term has since been used vigorously and often interchangeably with other terms such as citizen-generated spatial data (Mooney et al., 2011), neogeographic mapping (Turner, 2006), crowdsourced mapping (Miller & Goodchild, 2015), citizen sensing (Schade et al., 2013), and social sensing (Liu et al., 2015) to describe a collection of practices that involve collective mapping and spatial data generation by citizens.

Geo-participation is used in this book to refer to a collection of practices that uses geospatial information and technologies (e.g., maps and location-aware devices and services) and have participatory potential (Kar et al., 2016). The geo-participation practices and citizen-generated spatial data present opportunities for governments to embark on new methods of delivering public services, collecting citizen feedback broadening public participation and spur citizen collaboration and innovation (Degbelo et al., 2016; Sieber & Johnson, 2015). However, there have not been clear roadmaps for local governments to implement some of the new methods. Some of the geo-participation practices differ from PPGIS practices in terms of "geographic scale, data volume, and reach, as well the need for technological and cartographic expertise" (Sieber et al., 2016, p. 1033) and therefore require local governments to adopt different geospatial tools of collecting and managing data and to develop different participatory processes of spatial data generation.

Moreover, the multiplicity of terms generates confusion, as the lines between concepts and their associated practices are often blurred (See et al., 2016). For example, researchers have suggested that while PPGIS and VGI may involve similar spatial data collection processes, they often are used for distinct goals (i.e., PPGIS focuses on broadening participation, whereas VGI is primarily for spatial data generation) (Brown & Kyttä, 2014). It may not be clear to practitioners that how they may implement a geo-participation practice and how they may capitalize on a geo-participation practice to improve public participation. For researchers, the lack of a clear understanding of the association between geo-participation practices and their participation context may prevent researchers from examining relevant practices, assessing the usefulness of geo-participation for public participation in local government settings, and determining the research focus of one or a group of relevant geo-participation practices.

To this end, the first objective of this chapter is to propose a common typology that classifies geo-participation practices in the local government context. Geo-participation practices are classified based on their mechanisms of spatial data production and their potential uses for public participation in local government context. Based on the proposed typology, the second objective of this chapter is to identify the potential of emerging geo-participation practices for local governments to enhance public participation and to improve government operations. The identified opportunities have implications on the emerging needs of researching geo-participation.

3.2 Context: Geoweb and Geo-Participation

The initial efforts of incorporating GIS technologies into participatory public policy and planning process were motivated by the goals of bringing in local, experiential knowledge that often has a spatial component (e.g., important locations for a planning or development proposal) to decision-making processes and facilitating a wider scale participation (Brown & Kyttä, 2014; Kingston, 2007; Sieber, 2006). Digital mapping methods are primarily used in PPGIS projects for citizens to "map alternate views of the same problem and analyze the same data differently from those with political power" (Kar et al., 2016, p. 296). Government agencies primarily implement PPGIS projects to enhance public involvement and fulfill specific participation goals, such as engaging with marginalized groups, improving social inclusion, and building social capital (Brown & Kyttä, 2014). Participation, accordingly, is purposive, agency-driven, mostly small-scaled, and often involves deliberative activities (Brown, 2012; Sieber et al., 2016).

Developments in the Geoweb, a geographical version of Web 2.0, have expanded this view of participation. Web 2.0 is characterized by ubiquitous content generation by Internet users as a form of participation (O'Reilly, 2005). In the geographical context, the Geoweb technologies (i.e., a collection of online geospatial technologies that support online generation, dissemination, and management of geospatial information) have reduced technical barriers for lay people to create maps and have spurred a variety of activities of generating geographic information that is intrinsically participatory (Sieber et al., 2016). The wider scale spatial data generation has greatly expanded the outreach of PPGIS systems, as people now have fewer physical (e.g., space and time) and technical barriers to using online mapping applications (Tang & Liu, 2015). Moreover, new forms of participation that are intrinsic to the Geoweb have emerged and have brought a paradigm shift of how citizens engage with civic issues and policy-making (Elwood & Leszczynski, 2013). According to Sieber et al. (2016), participation that is intrinsic to the Geoweb differs from traditional PPGIS practices as it is often low-cost, less time-consuming, larger-scaled, and involves less formal coordination.

In the government context, new forms of government–citizen collaboration have emerged accordingly. Mirroring the vision of Goodchild (2007), that citizens could act as sensors to revitalize costly expert-driven data systems and to collectively observe and respond to a phenomenon, citizens may collaborate with government to improve and augment public data and services (Brabham, 2009; Collins et al., 2013). For example, citizens may provide up-to-date information (e.g., snow, flooding and traffic) to government services (Linders, 2012). There is also a potential for governments to involve citizens in updating, correcting, and vetting government geospatial datasets (Johnson, 2016; Sieber & Johnson, 2015).

A new form of passive citizen sensing has also emerged because of abundant geospatial data generated by citizens when they use location-aware devices and online location-based services. While the information is not generated with a participatory intention, it can be used to obtain aggregated local spatial knowledge related to time-critical situations (e.g., flood and earthquake), lived experiences (e.g., activity space and tourism experience), and public perceptions (e.g., public sentiment and perceived environment) (Cranshaw et al, 2012; Crooks et al., 2013; de Albuquerque et al., 2015; Dunkel, 2016; Zhou et al., 2015). In the local government context, passive citizen sensing has the potential to supplement or augment existing public participation approaches. Gao et al. (2017), for instance, suggest that public perceptions extracted from geotagged social media are comparable to those collected using in-person techniques.

The implementation of some of the new practices require local governments to adopt geospatial tools and design participation mechanisms that differ significantly from those of PPGIS. For example, passive citizen sensing may not involve the process of engaging with the public but requires local governments to deploy computational methods for extracting and analyzing user-generated data (Bright & Margetts, 2016). Some of the collaborative mapping practices may use digital mapping methods similar to those that are often used in PPGIS projects but aim to assembling data rather than to fulfill participation goals (e.g., widening participation and engaging with socially excluded groups) that are often addressed by PPGIS projects (Brown, 2017). Accordingly, the design and implementation of a collaborative mapping project may differently focus on spatial data collection rather than public engagement.

3.3 Developing a Typology for Geo-Participation

The Geoweb has spurred new forms of participation that differ from PPGIS. Participation that is on or intrinsic to the Geoweb can further be classified based on how information is generated by Internet users. According to Craglia et al. (2012), there are two ways of generating geographic information on the Geoweb (i.e., VGI). One is active sensing where users are explicitly volunteered to contribute specific information required by prior guidelines following pre-designed frameworks. The other is passive sensing where users are implicitly volunteered to contribute geospatial information usually with little prior guidance or as a by-product of other activities (e.g., using social media, making transactions, and using navigation apps).

In the government context, the two ways of generating geographic information associate with different uses of the information. Active sensing often relates to large-scaled practices of citizens actively contributing information for short-term goals, such as communicating service and infrastructure deficiency and reporting issues within spatial datasets. For example, citizens may directly report issues about public infrastructure and services and their locations to the government using location-enabled applications. Passive sensing may be used by local governments to obtain insights on public needs, lived experience, and human–environment interactions from user-generated spatial data (Liu et al., 2015).

Combining the differentiation of PPGIS and new forms of participation spurred by the Geoweb as well as the differentiation of active and passive sensing on the Geoweb, Table 3.1 describes a new typology for geo-participation. Consultative geo-participation refers to participatory practices that are aligned with traditional practices of PPGIS. While PPGIS is perhaps the most used term for participatory approaches with GIS components, there are other terms such as participatory GIS (PGIS), facilitated VGI (f-VGI), geo-questionnaire, and neogeogaphic mapping that brought by practitioners from various background to the field of PPGIS (Brown, 2017). Transactional geo-participation refers to active sensing on the Geoweb and often relates to government data and service delivery. Examples of transactional geo-participation include citizen reporting location-based civic issues to governments and citizen collaboration on collecting spatial data that can be used for government data and services. Passive geo-participation refers to passive sensing on the Geoweb and relates to a variety of terms (e.g., social sensing, ambient VGI, passive VGI) that describes the use of data-driven approaches to obtain collective perceptions and observations of real-world situations.

Table 3.1 A classification of geo-participation methods

Classification criteria		Geo-participation methods	Related terms and practices
Participation that utilizes GIS/the Geoweb	Active	Consultative Geo-Participation	PPGIS (Sieber, 2006); PGIS (Harris & Weiner, 1998); Facilitated VGI (f-VGI; Tulloch, 2008); Geo-questionnaire (Jankowski et al., 2016); Neogeographic mapping (Lin, 2013)
Participation that is intrinsic to the Geoweb	Active	Transactional Geo-Participation	Civic issue tracker (Sieber & Johnson, 2015); Participatory open data (Sieber & Johnson, 2015); OpenStreetMap
	Passive	Passive Geo-Participation	Social sensing (Liu et al., 2015); Ambient VGI (Stefanidis et al., 2013); Passive VGI (Craglia et al., 2012)

3.3.1 Consultative Geo-Participation

The aim of consultative geo-participation is to bring together multiple viewpoints and assist decision-making through "careful and serious weighing of reasons for and against some proposition" (Fearon, 1998, p. 63). The projects are usually initiated, guided and facilitated by the officials and are developed "as part of a pre-established planning or design process" (Seeger, 2008, p. 200). Citizen participants are primarily motivated by a "tendency to take part in activities that contribute to society" to participate (O'Brien et al., 2017, p. 321). Ideally, a participation project seeks to achieve a high level of citizen engagement where citizens may have the power to influence the processes and outcomes of decision-making (Aladalah et al., 2015). Outside the traditional arena of public participation, the Geoweb has spurred grassroots projects of citizen initiating their own mapping applications (i.e., neogeographic mapping) to discuss place-based issues. Although the grassroots projects may not directly impact decision-making, they create a virtual participatory space for citizens to give their voices and make the voices from below to become more visible (Lin, 2013; Sieber et al., 2016).

Digital mapping applications are often used in consultative geo-participation projects to collect local spatial knowledge regarding public perceptions of local environments, and participants are actively involved with commenting and discussing place-based issues regarding specific planning and development projects (Brown, 2012; Sieber, 2006). Particularly, visualization, sketching, communication, and evaluation functions are often included in digital mapping applications for participants to view spatial information, sketch, and comment on important locations and places, and reflect on each other's viewpoints and have in-depth discussions about place-based issues (Bugs et al., 2010; Carver et al., 2001; Kingston et al., 2000). With developments in the Geoweb, web-based digital mapping projects are increasingly used in consultative geo-participation projects to reduce the time and geographical constraints for citizens to participate.

Geospatial data collected from consultative participatory projects are mostly important locations and public perceptions related to specific planning and development projects (Brown, 2012). To what extent can these data affect decision-making outcomes, however, largely depends on whether and how governments accept processes and results of public participation (Brown, 2012). Less-controlled participation procedure and subsequently unassured quality of user-generated spatial data are identified as major barriers to local government adoption of geo-participation (Sieber, 2006). Besides, organizational constraints of public agencies being more likely to trust their own capabilities of influencing the public "from the inside through pressure politics and their own technical expertise" may also obscure the incorporation of citizen contributed spatial data (Brown, 2012, p. 15).

3.3.2 Transactional Geo-Participation

Transactional geo-participation intersects with consultative geo-participation in terms of active citizen contribution of spatial data and officially guided and facilitated data collection. Unlike consultative geo-participation projects that often involve deliberations on a particular planning or development project, transactional geo-participation is often used for improving government data and services. One example of transactional geo-participation is "Municipal 311," a program for issue reporting and service monitoring by municipal governments in North America (Nam & Pardo, 2014). Mostly following specific reporting procedures and prescribed to particular categories of civic issues, collective citizen reporting generates large amounts of geographic data that are mostly structured, and geographically and topically bounded (Sieber & Johnson, 2015). Government officials can use these data to further understand the geography of common issues and optimize service maintenance planning (O'Brien et al., 2017; Wiseman, 2014). Another instance of transactional geo-participation is the emerging practice of spurring citizen collaboration for correcting, updating, and editing government open geospatial data (Johnson, 2016). According to Sieber and Johnson (2015), this collaboration can be considered as a participatory model where data becomes a conduit for integrating citizen contribution into government operations and decision-making.

Two types of geospatial methods and techniques are primarily used for transactional geo-participation. The first are location-based applications and services. Civic issue tracking, for example, has been expanded from telephone-based reporting to web and mobile based reporting applications (Sieber & Johnson, 2015). Geotagging or geocoding functions are usually embedded in these apps for users to add location information. In some municipalities, reports of civic issues are further displayed using web maps to allow citizens to query and track reports.[1] With emerging practices such as participatory open data (i.e., citizens collaboratively update and vet government open data), mapping, and groupware tools are necessary for citizens to collaboratively contribute and edit data (e.g., tools of managing multiple contributions or versions for the same data). For local governments to enable such citizen collaboration, it is also necessary to develop proper mechanisms (e.g., establishing proper workflows of integrating citizen data and determining how much control the officials may possess for the data vetting process) to assure the fitness of citizen collaboration with government procedures (Johnson, 2016).

3.3.3 Passive Geo-Participation

Passive geo-participation deviates from traditional participation theories that consider public participation as a process that involves active participants (Rowe & Frewer, 2000). While some argue that citizens might be passive in public participation

[1] Cf: https://data.edmonton.ca/apps/311explorer/

process, they refer to citizens being receiving information from governments rather than actively contributing opinions and knowledge. Passive geo-participation differently emphasizes the unconsciousness and indirect citizen involvement and adopts data-driven approaches to extract public needs and opinions from citizen-generated data (Bright & Margetts, 2016).

In the context of government, passive geo-participation is only an emerging practice that has not been widely adopted and assessed. Two types of user-generated spatial data can be of potential uses for local governments to understand public needs and perceptions. The first is digital footprints of human activities (e.g., social media check-in data, taxi trajectories, cell phone positioning data), which are used to understand human-demarcated social areas, functional zones, commuting patterns, and spatial interaction (Cranshaw et al., 2012; Crooks et al., 2015; Liu et al., 2014). The second is geolocated qualitative data (e.g., geotagged short text and images, online articles, and blogs that have implicit geographic information) that are mixture of spatial information and multimedia content including text, image, video, and sound. These data are considered as, to some extent, reflections of lived experience and are used to explore collective perceptions of places, identify important locations of communal concern, and understand patterns of agreement and disagreement (Crooks et al., 2015; Dunkel, 2015).

To interpret big geospatial data and understand spatiotemporal patterns of human behaviors, developments of spatiotemporal analysis and visualization methods, and geoprocessing frameworks that account for veracity, velocity, and volume of big data are needed (Zhou et al., 2015). The increasing amounts of qualitative data and network-based data (e.g., communication networks) also call for the needs of combining text, image, and social network analysis methods with GIS systems to shed light on spatial expressions of public perceptions and social-spatial processes underpinning digital communication and interaction (Andris, 2016). Accordingly, the focus of deploying geospatial technology for passive geo-participation should be placed on collecting, processing, analyzing, and managing user-generated data.

3.4 Applying the Typology to Geo-Participation Literature

To demonstrate the relevance of the typology, this section applies the proposed typology to scholarly literature that present empirical works related to geo-participation from year 1995 to 2017. A comprehensive search was carried out on academic databases Web of Science and Scopus using searching terms PPGIS, PGIS, participatory GIS, VGI, f-VGI, neogeographic mapping, neogeography, crowdsourced mapping, crowdsourcing and GIS, crowdsourcing and planning, participatory and open data, ambient and geospatial planning, ambient and geospatial and decision-making. After removing duplicates in two databases, 2513 articles that contain at least one of aforementioned combination of keywords in their titles, abstracts, or keyword lists, were retrieved. A manual screening of titles and abstracts of these articles was then implemented to exclude irrelevant studies (e.g., studies that do not have a geospatial or participation component) and non-empirical studies (e.g., conceptual and theoretical studies, review papers, and tool developments). Note that some of the works included in the following analysis may indicate innovations in geospatial tools for public participation, yet they also present empirical case studies that provide evidence on public participation. Additionally, although this book is set in the context of governments, the selected scholar literature are not government studies only given the consideration that public participation mechanisms used in governments and public agencies may overlap.

In the remaining 157 articles, 24 articles are selected for further examination and classification using proposed typology (Table 3.2). These articles are citied the most (citation number no less than 15).[2] In addition, only the most cited article is included if multiple articles are presented in one subject of application (e.g., among five highly cited articles regarding consultative geo-participation for marine spatial planning, only the one with highest citation number is used). In the table, the enabling geospatial technologies and the research focus of each type of geo-participation are summarized based on identified articles.

The results show the conformity of studies within each category of geo-participation in terms of their research focuses. The results also suggest significant differences in techniques and processes used for three geo-participation practices, which indicate the validity of the proposed typology (Rowe & Frewer, 2005). Consultative geo-participation is applied to a variety of disciplines and relies heavily on participatory mapping and commenting platforms. An important emphasis of consultative geo-participation studies is the implications of mapping technologies on citizen empowerment, digital divide, and social inclusion. In particular, mapping practices are considered as complex social practices embedded with powers that may affect how and what local spatial knowledge is produced (Parker, 2006; Pickles, 2004). Transactional geo-participation initiatives are highly reliant on mobile technologies, location-based services, and groupware tools for crowdsourcing and collaboration. Studies in this category are often intertwined with citizens co-producing public services with governments and are evaluated by how the initiatives may improve the efficiency of public data and service delivery. Passive geo-participation initiatives mostly focus on development of methods for analyzing and interpreting user-generated spatial data. Not surprisingly, little scholarly literature has reported empirical evidence regarding how passive geo-participation may fit into government procedures. Technological developments are still heavily emphasized given the challenges of processing, analyzing, and managing the so-called big data. Nevertheless, a growing body of studies critically examines practices of adopting big data (Housley et al., 2014; Zook, 2017). These studies may shed light on further understandings of passive geo-participation initiatives.

[2] Note that different databases may differently count how many times an article is cited. For example, the article "The relationship between place attachment and landscape values: Toward mapping place attachment" (Brown & Raymond, 2007) was cited 225 times according to Scopus and 201 times according to Web of Science. To keep the consistency, this study uses citation number based on statistics of Scopus if an article is included in both Scopus and Web of Science databases.

Table 3.2 Geo-participation studies classified using the typology

Category	Examples	Summary	
		Geo	Participation
Consultative Geo-participation	• Evaluating values of public lands using PPGIS (Brown & Raymond, 2007) • Participatory mapping of cultural ecosystem services (Plieninger et al., 2013) • Participatory GIS for neighborhood planning (Al-Kodmany, 1999); marine spatial planning (St. Martin & Hall-Arber, 2008); national park planning (Brown & Weber, 2011); forest planning (Brown, 2009); wind energy planning (Simão et al., 2009); land use suitablity (Higgs, 2006); land use needs (Sandström et al., 2003); air pollution (Yearley et al., 2003); urban densification (Kyttä et al., 2013) • Map-based online forums for university master plan (Rinner, Keßler, & Andrulis, 2008) • Reach the hard-to-reach (Cinderby, 2010) Neographic mapping of Donghu, China (Lin, 2013)	• Digital mapping is primarily used in selected examples. In particular, map visualization and sketching functions are widely used	• The applications of consultative geo-participation encompass a wide range of fields • The participation outcome is strongly emphasized in the scholarly literature. Issues including how technology uses may enhance citizen participation, empower (or disempower) marginal groups, inducing digital divides are widely discussed
Transactional geo-participation	• Crowdsourcing through mobile phone applications for noise mapping (Rana et al., 2010); road safety (Aubry et al., 2014) • Social media reporting for soil and water conservation (Werts et al., 2012) VGI for cadastral mapping (Basiouka & Potsiou, 2012); transportation data collection (Misra et al., 2014)	• Geo-enabled instruments for transactional geo-participation include location-aware mobile devices and applications, social media, digital mapping • Spatial data quality is a significant challenge facing public agencies to adopt contributed data	• Transactional geo-participation is often applied to situations that are time-critical or lack accurate/ up-to-date data • The emphasis of scholar literature mostly centers upon the improvement of data/service
Category	Examples	Summary	

(continued)

Table 3.2 (continued)

Category	Examples	Summary	
		Geo	Participation
		Geo	Participation
Passive Geo-participation	• Real-time crisis mapping using social media (Middleton et al., 2014) • Mining location data to acquire spatiotemporal knowledge of forest fires (Longueville & Smith, 2009) • Crowdsourcing air temperature for energy planning (Overeem et al., 2013) • Understanding human-demarcated areas using social media data (Cranshaw et al., 2012) • Understanding perceived environment using crowdsourced photos (Dunkel, 2015)	• Passive geo-participation applications heavily rely on geospatial analytics • Commonly used geospatial methods in scholarly literature include: spatiotemporal analysis, geographic information retrieval (GIR), spatial data visualization, text analysis, topic modeling, spatial network analysis	• Selected examples are mostly experimental in that few have reported how these applications are used by practitioners • Emphasis of scholarly literature focuses primarily development of geospatial methodology, quality, and fitness-for-use of passively contributed data in specific application context

3.5 Emerging Opportunities for Local Government Operations

With developments in ICTs, promising new vehicles have emerged for local governments to better understand public needs, augment government services, improve government openness and accountability, and enhance citizen participation. In this section, we reflect on a collection of examples to shed light on the emerging opportunities presented by geo-participation for local government operations.

3.5.1 Going Beyond Transactional Geo-Participation

The increasing usage of smartphones and location-based services allow citizens to report emergencies and routine government services in a timely manner. According to Clark et al. (2013, p. 697), the wide scale citizen reporting enabled by smartphone application have "transformed citizens into "sensors," "detectors," or "reporters" to the problems facing the city." Ideally, citizens may collaborate via social networking functions and co-produce public services that are tailored to individual needs of citizens (Linders, 2012).

Figure 3.1 depicts a proof-of-concept of citizens collectively sharing information specific to pedestrian mobility. Citizen-generated information can be incorporated into the routing service that considers important factors (e.g., elevation, sidewalk width, surface materials) affecting pedestrians in order to help pedestrians find proper routes that suit their mobility. While road and sidewalk conditions are critical for people's daily commute, it is often challenging to rely solely on governments to update the dynamic situation in a timely manner (Rice et al., 2016). When citizens report a potential trip hazard to governments, this information is not immediately accessible to other citizens who need the information. The collective efforts, in contrast, may alleviate the challenges of timely updating sidewalk conditions (Neis & Zielstra, 2014). Moreover, multiple observations from the broader public have the potential of offering personalized services to accommodate various public needs. Specialized datasets of accessible or inaccessible features for disabled people is often lacking, not to mention up-to-date routing information that is tailored to disabled people (Prandi et al., 2014; Rice et al., 2016). Collectively sharing of information specific to pedestrian mobility (e.g., "intersection is not accessible for wheelchair users") can help citizens with varying needs to maneuver changing urban environment where transient barriers (e.g., crowds, temporary obstruction, and surface issues) often occur (Neis & Zielstra, 2014).

3.5.2 Passive Geo-Participation as an Aid to Understanding Local Spatial Knowledge

An ideal goal of utilizing passive geo-participation is to collect unfiltered views from broader segments of the society (Dunkel, 2015; Zhang & Feick, 2016). Figure 3.2 presents an example of using geotagged Twitter messages to collect public perceptions and needs related to a trail that is planned for improvement by the city. Iron Horse Trail is a landmark in the City of Kitchener, Canada. The trail provides local residents with unique outdoor experience and walking and cycling routes for commuting and exercising. Understanding and addressing public needs are important for city to develop improvement plan of the trail. From May 2016 to August 2016, 7789 geotagged Twitter messages were collected along the trail in the City of Kitchener using Twitter Streaming API.

Figure 3.2a shows the average numbers of geotagged Twitter messages collected within every hour at each day of the week. Compared to the pedestrians and cyclists counts at Iron trail based on the city's open data (Fig. 3.2b), a larger portion of Twitter messages were collected during the non-working hours before 8 am or after 6 pm, possibly due to the reduced likelihoods of people using Twitter during commuting.

To understand public's perceptions and preferences of urban space, sentiment analysis can be used to analyze tweets to obtain collective sentiments as indictors.

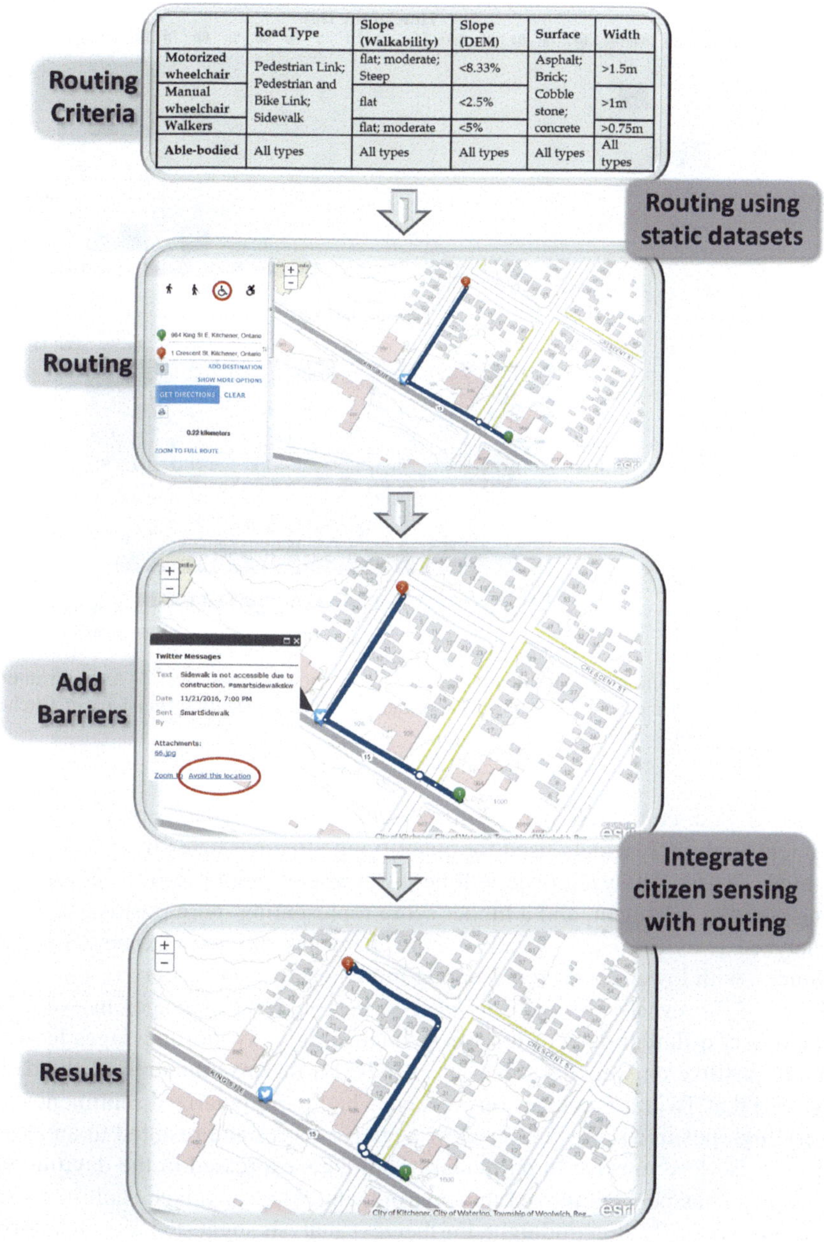

	Road Type	Slope (Walkability)	Slope (DEM)	Surface	Width
Motorized wheelchair	Pedestrian Link; Pedestrian and Bike Link; Sidewalk	flat; moderate; Steep	<8.33%	Asphalt; Brick; Cobble stone; concrete	>1.5m
Manual wheelchair		flat	<2.5%		>1m
Walkers		flat; moderate	<5%		>0.75m
Able-bodied	All types	All types	All types	All types	All types

Fig. 3.1 A citizen sensing approach to pedestrian route planning

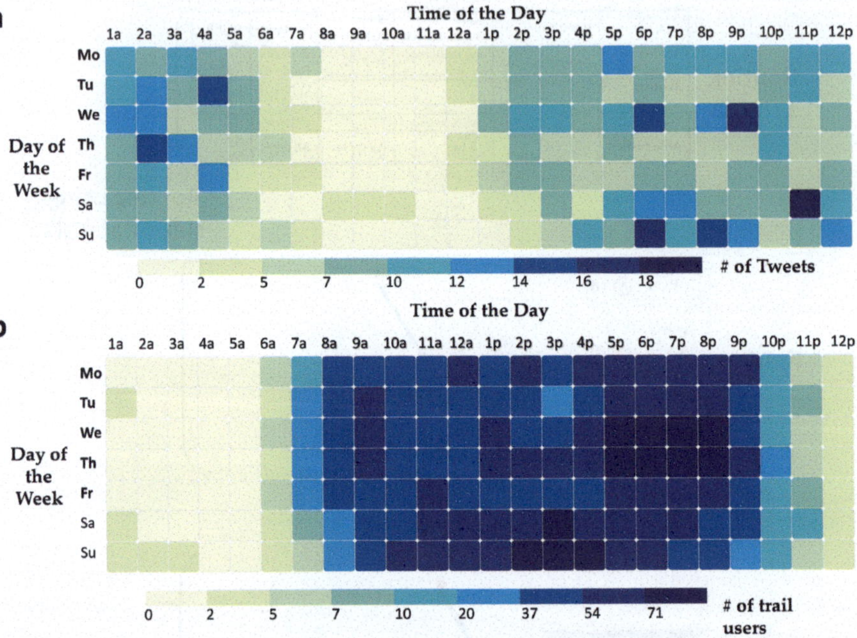

Fig. 3.2 (a) Average numbers of Twitter messages for every hour at weekdays; (b) Average numbers of trails users (pedestrians and cyclist) for every hour at weekdays. (*Data source* City of Kitchener open data catalogue)

Originated from computational linguistics and natural language analysis methods, sentiment analysis is widely used for quantifying affective states from texts (Zavattaro et al., 2015). Analyzed texts will be given a score with positive numbers indicating positive sentiment, and a higher score representing more positive sentiment. If analyzed texts receive a negative score, it means the texts express a negative sentiment, with lower scores representing more negative sentiment.

Figure 3.3a reveals average sentiment scores of tweet messages in every hour of the day at different days of the week. Interestingly, Twitter messages tended to be more positive on weekdays than weekends. Further examination of the dataset suggested that Twitter messages on weekends had more polarized sentiment scores of very high positive scores or very low negative scores and resulted in an average score that is close to zero. In particular, sentiments expressed in the daytime were more positive as people enjoyed the environment, whereas mixed sentiments were expressed especially after midnight. Further text analysis suggested several keywords such as light, sign, and crossings that might relate to the design of the trail (Fig. 3.3b). Not having streetlights and signage at some segments of the trail were mentioned in the tweets as people raised safety concerns. Positive words, comparatively, were more personalized and were not associated with particularly frequently used words.

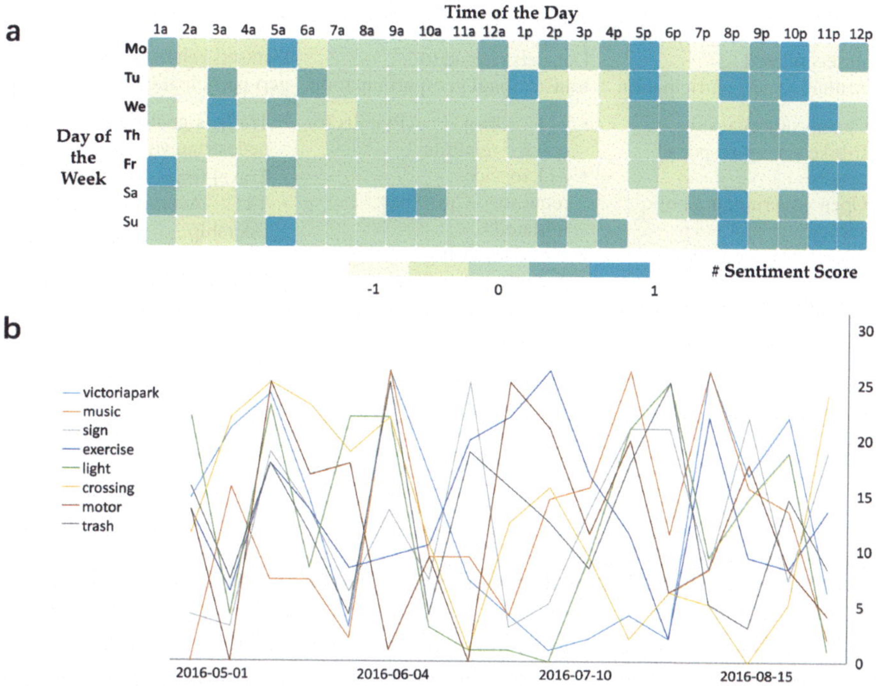

Fig. 3.3 (a) Average sentiment scores of Twitter messages for every hour at seven weekdays. (b) The frequency of the top words extracted from Twitter messages

Although not all of the Twitter messages were directly related to the surrounding environments, the sentiments expressed in the messages were mixed results of personal factors and people's emotional response to the places surrounding environment (Weinreb & Rofè, 2013). Using text analysis methods as presented in this example, decision-makers may obtain insights on the trending expressions and shared feelings that present unfiltered views of landscapes and surrounding environments.

3.5.3 Integration of Geo-Participation with Government Open Data Programs

Recently, there is growing trend of fostering greater openness of governments at different levels. As an important component of this effort, government open data programs serve as the basis of information sharing, as it allows the public to use and share government data (Open Knowledge, n.d.; Veljković et al., 2014). Given the significant amounts of open geospatial data, spatial technology plays an increasingly critical role for data sharing, visualization, and interpretation (Johnson et al., 2017). For instance, map interfaces are widely used by municipalities for visualizing and

Table 3.3 A list of Toronto open data sets related to geo-participation

Datasets related to consultative geo-participation	Datasets related to transactional geo-participation	Datasets related to passive geo-participation
• Casino survey results • E-bike survey response results • Open government public survey • Transform to community feedback • Woodbine Racetrack Casino consultation 2015 • Polls conducted by the city • Poverty reduction strategy consultation—Qualitative input • Street needs assessment results	• 311—Open311 API calls for service requests • 311 service requests—Customer initiated • Flood reporting noted by Toronto water districts	• Traffic signal vehicle and pedestrian volumes • Travel times—Bluetooth • TTC—Average weekday ridership

cataloguing open data. Citizen contributions are also becoming an important component of government open data processes. Sieber and Johnson (2015), for example, suggested that citizens can contribute data by reporting civic issues and collectively vetting government open data. Some citizen data coming from consultative and passive participation channels is also made public accessible.

Through a comprehensive review of City of Toronto's open data catalogue, we identified 14 datasets out of total 267 open datasets that are relevant to geo-participation (Table 3.3).[3] The selection criteria include: (1) the dataset must have geospatial attributes and (2) the dataset must involve with either active or passive public input. Among the 14 selected datasets, 8 datasets are related to consultative participation approaches, such as public survey, e-polling, and public consultation. Four are related to transactional geo-participation such as non-emergency (2 datasets) and emergency (1 dataset) reporting. The other three datasets are citizens' travel behaviors recorded by traffic cameras, Bluetooth, and Wi-Fi sensors, which are related to passive geo-participation.

Although there are more government programs related to geo-participation practices, datasets listed in Table 3.2 provide us with a glance of how geo-participation practices are closely related to government programs. Geospatial data resulted from participation practices can further support other forms of citizen participation such as app developments by independent developers or third-party companies and organizations (Kassen, 2013). For example, public survey results of developing a casino were visualized using map applications.[4] Outside of the government mandate, these

[3] The website was accessed on October 1, 2017.

[4] (a) Source: http://justinpierre.ca/casino/
(b) Source: http://laurenarcher.github.io/CasinoSurveyVis/

third-party applications help the public examine survey results and serve as intermediaries to provide the public with value-added information and assist with those who may not possess technical understandings of open data (Gagliardi et al., 2015). In this regard, while three geo-participation methods involve different techniques and processes, geospatial data generated from different geo-participation methods may be used across the three categories and facilitate governmental initiatives of collaborating with citizens.

3.6 Implications for Implementing and Researching Geo-Participation

To move forward with the emerging opportunities presented by geo-participation practices, further investigations are necessary to develop tools and techniques for implementing the emerging geo-participation practices and to understand the social implication of the practices.

3.6.1 Spatial Data Management

The practices of opening up citizen data and associated civic engagement activities as introduced in Sect. 3.5.3 have indications on producing, disseminating, and managing geospatial data within local governments (Fig. 3.4). Data collected from all geo-participation methods can be brought together to support decision-making, government operation, and service delivery. Some of these data may also be included in open data programs following open data standards. The release of these data can further facilitate citizen participation through app developments, which are conducive to enhancing citizen participation and augmenting government services. Similar to open data and innovation ecosystem suggested by Gagliardi et al. (2015), it is important to leverage data needs of both governments and citizens and generate mechanisms to provide feedback and initiate secondary use of geospatial data (i.e., use data for app development, produce public services, etc.).

At the core of this data ecosystem, spatial data infrastructure (SDI) should provide an environment for managing and using heterogeneous geospatial data. SDI refers to “Internet-based mechanisms for the coordinated production, discovery, and use of geospatial information in the digital environment” (Budhathoki et al., 2008 p.49). Traditional SDIs are based on the model of governments being publishers and citizens being data users (Hu & Li, 2017). SDIs for the data ecosystem needs to further consider seamlessly integrating spatial data from different sources, particularly data vetted and published by non-experts (i.e., citizens). The efforts of integrating input from citizen input into government SDIs are still in the early stages given the challenge of reconciling relevant yet heterogeneous datasets (Coleman et al., 2016).

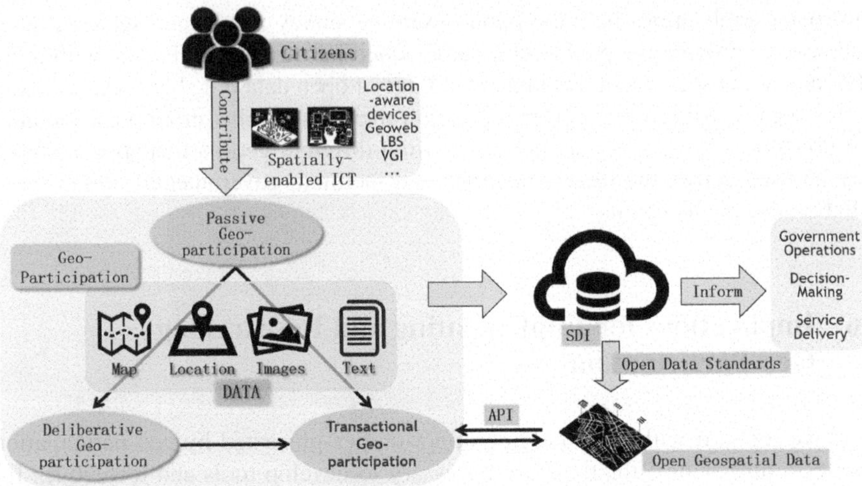

Fig. 3.4 A synthesized view of spatial data flows among geo-participation and government operations

Geospatial data may be created with various formats (e.g., user-generated maps, text-based comments and tags, media content such as images and videos) and by authors with varying geographic knowledge and diverging perceptions of on-the-ground situations (Bakillah et al., 2013). It is required to develop relevant technologies, standards, and polices to accommodate the influx of data with varying accuracy and currency (Hu & Li, 2017). To enable citizen collaboration as suggested in Sect. 3.5.1, governments also need to develop mechanisms that support data and service interoperability and to tackle the issues of liability and privacy raised by making data and services publicly available (Bakillah et al., 2013; Scassa & Diebel, 2017).

3.6.2 *Utilizing Passive Geo-Participation*

Passive geo-participation is an emerging practice that has the potential to help local governments harvest local spatial knowledge from citizen-generated data and enlarge public participation processes (Bright & Margetts, 2016). However, little evidence insofar has been provided in scholarly literature regarding applying findings from passive geo-participation to government decision-making. The potential of using passive geo-participation to solicit input from broader groups of citizens needs to be further assessed by addressing the following two questions (Bright & Margetts, 2016): first, what kinds of public opinions or perceptions regarding general or specific policy and decision-making can be harvested? Second, whom do the data represent? In other words, who are the participants? Answers to these questions are essential for evaluating passive geo-participation practices and understanding their impacts on participation outcomes.

From a spatial data perspective, incorporating passive geo-participation into government operations may face a number of challenges. First, lacking the trustworthiness of user-generated data, particularly relating to multiple data resources and lacking provenance information, can be major barriers for local government to integrate research findings with decision-making (Johnson, 2016). Second, exploitation of passive geo-participation is often associated with big spatial analytics for processing and analyzing unstructured data. In particular, increasing amounts of text, video stream, and image data are used by government as reported by Gagliardi et al. (2015). On the one hand, it is necessary to develop data-driven approaches such as geographic information retrieving (GIR) methods to extract geospatial information from non-spatial data (e.g., text descriptions, images, and videos) that include abundant local spatial knowledge (Vasardani et al., 2013). On the other hand, the heavy reliance on the technical capability to processing big geospatial data can also prevent local governments from using passive geo-participation (Zhang & Feick, 2016). Empirical studies are necessary to further investigate the challenge and to identify proper mechanisms of utilizing geo-participation in local government context.

3.6.3 Understanding the Social and Spatial Implications of Geo-Participation

One premise of utilizing the Geoweb for public participation is that the wider scale spatial data generation may broaden public participation and accordingly enable collective wisdom. However, there are continuing debates on whether the technological-driven methods may exacerbate existing social disparity or generate new digital divides (Sieber et al., 2016). For example, citizens will need access to Internet and digital devices (e.g., computers, mobile phones) to use the digital services. Empirical studies have suggested that factors including geography and sociodemographic characteristics differently affect how people participate via digital methods (Clark et al., 2013; Foster & Dunham, 2014).

It also has not been clear how would the broadened scale affect participation process. Localness is often considered as important to public participation, given that local communities are the primary sources of local knowledge and are more likely to be affected by decision outcomes (Sieber, 2006). The notion of localness also affects the trustworthiness of local spatial knowledge, as it is believed that local participants are more familiar with local circumstances and thus will provide more reliable and accurate information (Brabham, 2009; Brown & Kyttä, 2014). In the online environment, Wellman (2002) suggest that participation is composed of "intense local and extensive global interaction(s)" (Wellman, 2002, p. 11). Accordingly, the notion of local might have changed as group relationships within the community have become "glocalized." Taking social media, the increasingly popular platform of government–citizen communication, as an example, a city's official account may have a significant portion of followers that are not local residents. Using Twitter's Friends/Followers API, we collected self-claimed locations of the followers of City of Waterloo's official Twitter account and found that 13.1% of them were out of the country, and additional 22.7% outside of the city/region. While self-claimed locations do not always accurately reflect users' physical locations, previous studies confirmed the roots of online social networks in physical communities (Stephens & Poorthuis, 2015). This estimate can shed some light on the globalized composition of public that is often beyond the city's jurisdiction.

The interactions between governments and the public on social media, however, suggest more localized patterns. Figure 3.5 presents government–citizen communication network generated based on the Region of Waterloo and the cities of Waterloo and Kitchener's Twitter accounts. The size of a node indicates how frequent the account represented by the node interacts (e.g., reposting a post and have direct dialogues) with the region and cities' accounts. Compared to the number of followers of the region and cities' accounts, only a relatively small group of local business and organizations is actively involved with online conversations. The presence of some physically distant Twitter users who are actively engaged with online conversations has indications on considering the "localness" of participation to be not only physically close but also socially proximate. That is, one that is physically outside of the community but remains connected with local communities and active involved with local affairs should still be considered to be local. However, it has not been clear how this alternative view of localness may affect the trustworthiness of citizen-generated data and accordingly government–citizen collaboration.

It is necessary for researchers and practitioners to further examine the social and spatial processes underpinning geo-participation. As suggested by Clark et al. (2013), understanding how technologies affect citizen participation is critical to assess the benefits and limitations of technologically enabled government–citizen collaborations. GIS methods should be introduced to this investigation of social and spatial implications of geo-participation. Particularly, attention should be paid to exploring relationships between online and offline communities as well as associations of social and geographical connections. Traditional spatial models such as distance decay models may be useful for testing the assumption of distance effects on Internet-based participation. Methods such as social network analysis (SNA) may assist with exploration of the social dimension of localness. In fact, there is recent call for the integration of SNA into GIS systems to better understand the geography of communication and social interaction networks in GIS research (Andris, 2016). Synthesizing this research agenda with geo-participation context may advance our understanding of how the emerging geo-participation practices affect citizen participation.

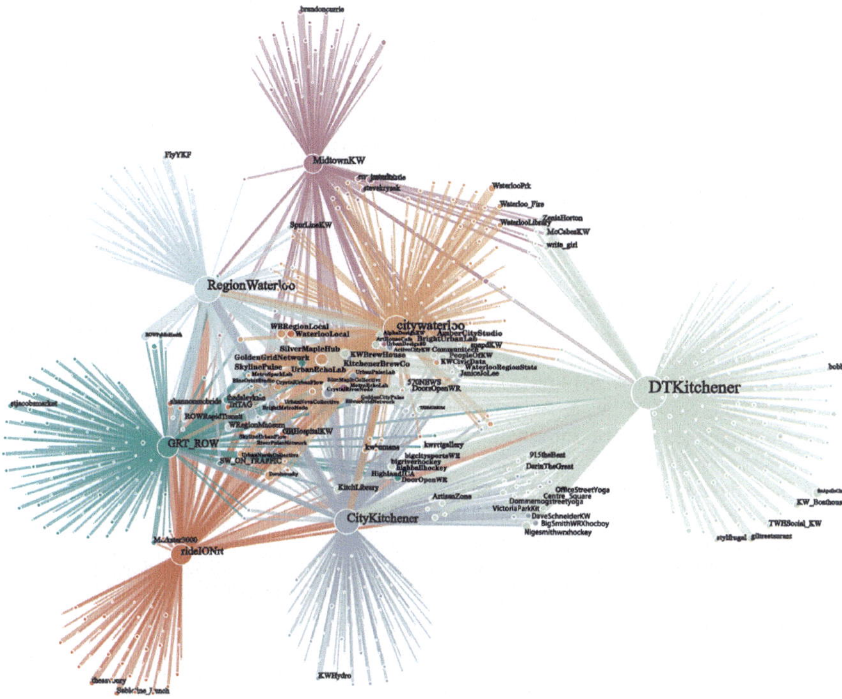

Fig. 3.5 Networks of Region of Waterloo and cities of Kitchener and Waterloo's Twitter conversation

3.7 Conclusion

The continued developments in ICTs are conducive to the proliferation of user-generated spatial data and new forms of geo-participation. To elucidate the diffusion of diverse concepts and practices related to geo-participation, this chapter proposes an innovative typology for examining and implementing geo-participation. The intent of this typology is to clarify techniques, processes, and outcomes associated with different geo-participation practices and establish connections between geo-participation and government operations. The typology along with examples demonstrated in the chapter suggests emerging geo-participation practices that are conducive to enhanced public participation and improved government–citizen collaboration. Future research should further assess the identified potential and determine the best practices of utilizing emerging geo-participation practices in local government settings. More work is also needed to understand how the wider scale data generation affect public participation processes and change dynamics of government–citizen interactions.

References

Aladalah, M., Cheung, Y., & Lee, V. (2015). Enabling citizen participation in Gov 2.0: An Empowerment perspective. *Electronic Journal of E-Government, 13*(2), 77–93.

Al-Kodmany, K. (1999). Using visualization techniques for enhancing public participation in planning and design: Process, implementation, and evaluation. *Landscape and Urban Planning, 45*(1), 37–45. https://doi.org/10.1016/S0169-2046(99)00024-9

Andris, C. (2016). Integrating social network data into GISystems. *International Journal of Geographical Information Science, 8816*(March), 1–23. https://doi.org/10.1080/13658816.2016.1153103

Aubry, E., Silverston, T., Lahmadi, A., & Festor, O. (2014). CrowdOut: A mobile crowdsourcing service for road safety in digital cities. In *2014 IEEE International Conference on Pervasive Computing and Communication Workshops, PERCOM WORKSHOPS 2014* (pp. 86–91). https://doi.org/10.1109/PerComW.2014.6815170

Bakillah, M., Liang, S., Zipf, A., & Arsanjani, J. (2013). Semantic interoperability of sensor data with Volunteered geographic information: A unified model. *ISPRS International Journal of Geo-Information, 2,* 766–796. https://doi.org/10.3390/ijgi2030766

Basiouka, S., & Potsiou, C. (2012). VGI in cadastre: A Greek experiment to investigate the potential of crowd sourcing techniques in cadastral mapping. *Survey Review, 44*(325), 153–161. https://doi.org/10.1179/1752270611Y.0000000037

Brabham, D. C. (2009). Crowdsourcing the public participation process for planning projects. *Planning Theory, 8*(3), 242–262. https://doi.org/10.1177/1473095209104824

Bright, J., & Margetts, H. (2016). Big data and public policy: can it succeed where e-participation has failed? *Policy and Internet, 8*(3), 218–224.

Brown, G. (2009). Public participation GIS: A new method for use in national forest planning. *Forest Science, 55*(2), 166–182.

Brown, G. (2012). Public participation GIS (PPGIS) for regional and Environmental Planning: reflections on a decade of empirical research. *URISA Journal, 25*(2), 7–18.

Brown, G. (2017). A review of sampling effects and response bias in Internet participatory mapping (PPGIS/PGIs/VGI). *Transactions in GIS, 21*(1), 39–56. https://doi.org/10.1111/tgis.12207

Brown, G., & Raymond, C. (2007). The relationship between place attachment and landscape values: Toward mapping place attachment. *Applied Geography, 27*(2), 89–111. https://doi.org/10.1016/j.apgeog.2006.11.002

Brown, G., & Weber, D. (2011). Public participation GIS: A new method for national park planning. *Landscape and Urban Planning, 102*(1), 1–15. https://doi.org/10.1016/j.landurbplan.2011.03.003

Brown, G., & Kyttä, M. (2014). Key issues and research priorities for public participation GIS (PPGIS): A synthesis based on empirical research. *Applied Geography, 46,* 126–136. https://doi.org/10.1016/j.apgeog.2013.11.004

Budhathoki, N. R., Bruce, B. (Chip), & Nedovic-Budic, Z. (2008). Reconceptualizing the role of the user of spatial data infrastructure. *GeoJournal, 72*(3–4), 149–160. https://doi.org/10.1007/s10708-008-9189-x

Bugs, G., Granell, C., Fonts, O., Huerta, J., & Painho, M. (2010). An assessment of public participation GIS and web 2.0 technologies in urban planning practice in Canela, Brazil. *Cities, 27*(3), 172–181. https://doi.org/10.1016/j.cities.2009.11.008

Carver, S., Evans, A., Kingston, R., & Turton, I. (2001). Public participation, GIS, and cyberdemocracy: Evaluating on-line spatial decision support systems. *Environment and Planning b: Planning and Design, 28*(6), 907–921. https://doi.org/10.1068/b2751t

Cinderby, S. (2010). How to reach the 'hard-to-reach': The development of participatory geographic information systems (P-GIS) for inclusive urban design in UK cities. *Area, 42*(2), 239–251. https://doi.org/10.1111/j.1475-4762.2009.00912.x

Clark, B. Y., Brudney, J. L., Jakobsen, M., & Andersen, S. C. (2013). Coproduction of government services and the new information technology: Investigating the distributional biases. *Public Administration Review, 73*(5), 704–713. https://doi.org/10.1111/puar.12092.Coproduction

Coleman, D. J., Rajabifard, A., & Kolodziej, K. W. (2016). Expanding the SDI environment: Comparing current spatial data infrastructure with emerging indoor location-based services. *International Journal of Digital Earth, 8947*(January), 1–19. https://doi.org/10.1080/17538947.2015.1119207

Collins, L., Swart, R., & Zhang, B. (2013). *The rise of future finance: The UK alternative finance benchmarking report*. Nesta; University of Cambridge.

Craglia, M., Ostermann, F., & Spinsanti, L. (2012). Digital Earth from vision to practice: Making sense of citizen-generated content. *International Journal of Digital Earth, 5*(5), 398–416. https://doi.org/10.1080/17538947.2012.712273

Cranshaw, J., Hong, J. I., & Sadeh, N. (2012). The Livehoods project: Utilizing social media to understand the dynamics of a city. In *Proceedings of the Sixth International AAAI Conference on Weblogs and Social Media* (pp. 58–65).

Crooks, A., Croitoru, A., Stefanidis, A., & Radzikowski, J. (2013). #Earthquake: Twitter as a distributed sensor system. *Transactions in GIS, 17*(1), 124–147. https://doi.org/10.1111/j.1467-9671.2012.01359.x

Crooks, A., Pfoser, D., Jenkins, A., Croitoru, A., Stefanidis, A., Smith, D., … Lamprianidis, G. (2015). Crowdsourcing urban form and function. *International Journal of Geographical Information Science, 29*(5), 720–741. https://doi.org/10.1080/13658816.2014.977905

de Albuquerque, J. P., Herfort, B., Brenning, A., & Zipf, A. (2015). A geographic approach for combining social media and authoritative data towards identifying useful information for disaster management. *International Journal of Geographical Information Science, 29*(4), 667–689. https://doi.org/10.1080/13658816.2014.996567

Degbelo, A., Granell Granell, C., Trilles Oliver, S., Bhattacharya, D., Casteleyn, S., & Kray, C. (2016). Opening up smart cities: citizen-centric challenges and opportunities from GIScience. *ISPRS International Journal of Geo-Information, 5*(2), 16. https://doi.org/10.3390/ijgi5020016

Dunkel, A. (2015). Visualizing the perceived environment using crowdsourced photo geodata. *Landscape and Urban Planning, 142,* 173–186. https://doi.org/10.1016/j.landurbplan.2015.02.022

Dunkel, A. (2016). *Assessing the perceived environment through crowdsourced spatial photo content for application to the fields of landscape and urban planning,* (August).

Elwood, S., & Leszczynski, A. (2013). New spatial media, new knowledge politics. *Transactions of the Institute of British Geographers, 38*(4), 544–559. https://doi.org/10.1111/j.1475-5661.2012.00543.x

Fearon, J. D. (1998). Deliberation as discussion. In *Deliberative Democracy* (pp. 44–68). https://doi.org/10.1017/CBO9781139175005.004

Foster, A., & Dunham, I. M. (2014). Volunteered geographic information, urban forests, & environmental justice. *Computers, Environment and Urban Systems, 53,* 65–75. https://doi.org/10.1016/j.compenvurbsys.2014.08.001

Gagliardi, D., Schina, L., Sarcinella, M. L., Mangialardi, G., Niglia, F., & Corallo, A. (2015). Information and communication technologies and public participation: Interactive maps and value added for citizens. *Government Information Quarterly, 34*(1), 153–166. https://doi.org/10.1016/j.giq.2016.09.002

Gao, S., Janowicz, K., Montello, D. R., Hu, Y., Yang, J., McKenzie, G., Ju, Y., & Yan, B. (2017). A data-synthesis-driven method for detecting and extracting vague cognitive regions. *International Journal of Geographical Information Science, 31*(6). https://doi.org/10.1080/13658816.2016.1273357

Goodchild, M. F. (2007). Citizens as sensors: The world of volunteered geography. *GeoJournal, 69*(4), 211–221. https://doi.org/10.1007/s10708-007-9111-y

Harris, T., & Weiner, D. (1998). Empowerment, marginalization, and "Community-integrated" GIS. *Cartography and Geographic Information Systems, 25*(2), 67–76. https://doi.org/10.1559/152304098782594580

Higgs, G. (2006). Integrating multi-criteria techniques with geographical information systems in waste facility location to enhance public participation. *Waste Management and Research, 24*(2), 105–117. https://doi.org/10.1177/0734242X06063817

Housley, W., Procter, R., Edwards, A., Burnap, P., Williams, M., Sloan, L., Rana, O., Morgan, J., Voss, A., & Greenhill, A. (2014). Big and broad social data and the sociological imagination: A collaborative response. *Big Data & Society, 1*(2), 1–15. https://doi.org/10.1177/2053951714545135

Hu, Y., & Li, W. (2017). Spatial Data Infrastructure. In J. P. Wilson (Ed.), *The Geographic Information Science & Technology Body of knowledge*. https://doi.org/10.22224/gistbok/2017.2.1

IAP2. (n.d.). *Public participation spectrum*. Retrieved September 8, 2017, from http://iap2canada.ca/page-1020549

Jankowski, P., Czepkiewicz, M., Młodkowski, M., & Zwoliński, Z. (2016). Geo-questionnaire: A method and tool for public preference elicitation in land use planning. *Transactions in GIS, 20*(6), 903–924. https://doi.org/10.1111/tgis.12191

Johnson, P. A. (2016). Models of direct editing of government spatial data: Challenges and constraints to the acceptance of contributed data. *Cartography and Geographic Information Science, 0406*(May), 1–11. https://doi.org/10.1080/15230406.2016.1176536

Johnson, P. A., Sieber, R. E., Scassa, T., Stephens, M., & Robinson, P. (2017). The cost(s) of geospatial open data. *Transactions in GIS, 21*(3), 434–445. https://doi.org/10.1111/tgis.12283

Kar, B., Sieber, R. E., Haklay, M., & Ghose, R. (2016). Public participation GIS and participatory GIS in the era of GeoWeb. *The Cartographic Journal, 53*(4), 296–299. https://doi.org/10.1080/00087041.2016.1256963

Kassen, M. (2013). A promising phenomenon of open data: A case study of the Chicago open data project. *Government Information Quarterly, 30*(4), 508–513. https://doi.org/10.1016/j.giq.2013.05.012

Kingston, R. (2007). Public participation in local policy decision-making: The role of web-based mapping. *The Cartographic Journal, 44*(2), 138–144. https://doi.org/10.1179/000870407X213459

Kingston, R., Carver, S., Evans, A., & Turton, I. (2000). Web-based public participation geographical information systems: An aid to local environmental decision-making. *Computers, Environment and Urban Systems, 24*(2), 109–125. https://doi.org/10.1016/S0198-9715(99)00049-6

Kyttä, M., Broberg, A., Tzoulas, T., & Snabb, K. (2013). Towards contextually sensitive urban densification: Location-based softGIS knowledge revealing perceived residential environmental quality. *Landscape and Urban Planning, 113,* 30–46. https://doi.org/10.1016/j.landurbplan.2013.01.008

Lin, W. (2013). Situating performative neogeography: Tracing, mapping, and performing "Everyone's East Lake." *Environment and Planning A, 45*(1), 37–54. https://doi.org/10.1068/a45161

Linders, D. (2012). From e-government to we-government: Defining a typology for citizen coproduction in the age of social media. *Government Information Quarterly, 29*(4), 446–454. https://doi.org/10.1016/j.giq.2012.06.003

Liu, Y., Sui, Z., Kang, C., & Gao, Y. (2014). Uncovering patterns of inter-urban trip and spatial interaction from social media check-in data. *PLoS ONE, 9*(1), Article e86026. https://doi.org/10.1371/journal.pone.0086026

Liu, Y., Liu, X., Gao, S., Gong, L., Kang, C., Zhi, Y., Chi, G., & Shi, L. (2015). Social sensing: A new approach to understanding our socioeconomic environments. *Annals of the Association of American Geographers, 105*(3), 512–530. https://doi.org/10.1080/00045608.2015.1018773

Longueville, B. De, & Smith, R. S. (2009). "OMG, from here, I can see the flames !": a use case of mining Location Based Social Networks to acquire spatio- temporal data on forest fires. In

Proceedings of the 2009 International Workshop on Location Based Social Networks (LBSN '09) (pp. 73–80). Association for Computing Machinery. https://doi.org/10.1145/1629890.1629907

Middleton, S. E., Middleton, L., & Modafferi, S. (2014). Real-time crisis mapping of natural disasters using social media. In *IEEE Intelligent Systems* (Vol. 29, pp. 9–17). https://doi.org/10.1109/MIS.2013.126

Miller, H. J., & Goodchild, M. F. (2015). Data-driven geography. *GeoJournal,* (October 2014), 449–461. https://doi.org/10.1007/s10708-014-9602-6

Misra, A., Gooze, A., Watkins, K., Asad, M., & Le Dantec, C. (2014). Crowdsourcing and Its Application to Transportation Data Collection and Management. *Transportation Research Record: Journal of the Transportation Research Board, 2414,* 1–8. https://doi.org/10.3141/2414-01

Mooney, P., Sun, H., Corcoran, P., & Yan, L. (2011). Citizen-generated spatial data and information: Risks and opportunities. *Proceedings of 2011 IEEE International Conference on Intelligence and Security Informatics,* 232–232. https://doi.org/10.1109/ISI.2011.5984087

Nam, T., & Pardo, T. A. (2014). The changing face of a city government: A case study of Philly311. *Government Information Quarterly, 31*(SUPPL.1), S1–S9. https://doi.org/10.1016/j.giq.2014.01.002

Neis, P., & Zielstra, D. (2014). Recent developments and future trends in volunteered geographic information research: The case of OpenStreetMap. *Future Internet, 6*(1), 76–106. https://doi.org/10.3390/fi6010076

O'Brien, D. T., Offenhuber, D., Baldwin-Philippi, J., Sands, M., & Gordon, E. (2017). Uncharted territoriality in coproduction: the Motivations for 311 reporting. *Journal of Public Administration Research and Theory,* 320–335. https://doi.org/10.1093/jopart/muw046

Open Knowledge. (n.d.). *Open definition.* Retrieved March 14, 2018, from http://opendefinition.org/

O'Reilly, T. (2005). *O'Reilly: What Is Web 2.0.* Retrieved from http://www.oreillynet.com/pub/a/oreilly/tim/news/2005/09/30/what-is-web-20.html

Overeem, A., R. Robinson, J. C., Leijnse, H., Steeneveld, G. J., P. Horn, B. K., & Uijlenhoet, R. (2013). Crowdsourcing urban air temperatures from smartphone battery temperatures. *Geophysical Research Letters, 40*(15), 4081–4085. https://doi.org/10.1002/grl.50786

Parker, B. (2006). Constructing community through maps? Power and praxis in community mapping. *Professional Geographer, 58*(4), 470–484. https://doi.org/10.1111/j.1467-9272.2006.00583.x

Pickles, J. (2004). *A history of Spaces: Cartographic Reason, Mapping and the Geo-Coded World.* Routledge.

Plieninger, T., Dijks, S., Oteros-Rozas, E., & Bieling, C. (2013). Assessing, mapping, and quantifying cultural ecosystem services at community level. *Land Use Policy, 33,* 118–129. https://doi.org/10.1016/j.landusepol.2012.12.013

Prandi, F., Soave, M., Devigili, F., & Amicis, R. De. (2014). *Collaboratively collected geodata to support routing service for disabled people,* (November), 26–28.

Rana, R. K., Chou, C. T., Kanhere, S. S., Bulusu, N., & Hu, W. (2010). Ear-phone: An end-to-end participatory urban noise mapping system. *Proceedings of the International Conference on Information Processing in Sensor Networks IPSN,* 105–116. https://doi.org/10.1145/1791212.1791226

Rice, R. M., Aburizaiza, A. O., Rice, M. T., & Qin, H. (2016). Position validation in crowdsourced accessibility mapping. *Cartographica: The International Journal for Geographic Information and Geovisualization, 51*(2), 55–66. https://doi.org/10.3138/cart.51.2.3143

Rinner, C., Keßler, C., & Andrulis, S. (2008). The use of Web 2.0 concepts to support deliberation in spatial decision-making. *Computers, Environment and Urban Systems, 32*(5), 386–395. https://doi.org/10.1016/j.compenvurbsys.2008.08.004

Rowe, G., & Frewer, L. J. (2000). Public Participation Methods: A Framework for Evaluation. *Science, Technology & Human Values, 25*(1), 3–29. https://doi.org/10.1177/016224390002500101

Rowe, G., & Frewer, L. J. (2005). A typology of public engagement mechanisms. *Science, Technology & Human Values, 30*(2), 251–290. https://doi.org/10.1177/0162243904271724

Sandström, P., Granqvist Pahlén, T., Edenius, L., Tømmervik, H., Hagner, O., Hemberg, L., Olsson, H., Baer, K., Stenlund, T., Brandt, L. G., & Egberth, M. (2003). Conflict resolution by participatory management: Remote sensing and GIS as tools for communicating land-use needs for reindeer herding in Northern Sweden. *AMBIO: A Journal of the Human Environment, 32*(8), 557–567. https://doi.org/10.1579/0044-7447-32.8.557

Scassa, T., & Diebel, A. (2017). Open or closed? Open licensing of real-time public sector transit data.

Schade, S., Díaz, L., Ostermann, F., Spinsanti, L., Luraschi, G., Cox, S., Nuñez, M., & De Longueville, B. (2013). Citizen-based sensing of crisis events: Sensor web enablement for volunteered geographic information. *Applied Geomatics, 5*(1), 3–18. https://doi.org/10.1007/s12518-011-0056-y

See, L., Mooney, P., Foody, G. M., Bastin, L., Comber, A., Estima, J., … Laakso, M. (2016). Crowdsourcing, citizen science or volunteered geographic information? *The Current State of Crowdsourced Geographic Information,* (April), 1–10. https://doi.org/10.3390/www.mdpi.com/journal/ijgi

Seeger, C. J. (2008). The role of facilitated volunteered geographic information in the landscape planning and site design process. *GeoJournal, 72*(3–4), 199–213. https://doi.org/10.1007/s10708-008-9184-2

Sieber, R. E. (2006). Public participation geographic information systems: A literature review and framework. *Annals of the Association of American Geographers, 96*(3), 491–507. https://doi.org/10.1111/j.1467-8306.2006.00702.x

Sieber, R. E., & Johnson, P. A. (2015). Civic open data at a crossroads: Dominant models and current challenges. *Government Information Quarterly, 32*(3), 308–315. https://doi.org/10.1016/j.giq.2015.05.003

Sieber, R. E., Robinson, P. J., Johnson, P. A., & Corbett, J. M. (2016). Doing public participation on the Geospatial Web. *Annals of the American Association of Geographers, 106*(5), 1030–1046. https://doi.org/10.1080/24694452.2016.1191325

Simão, A., Densham, P. J., & Haklay, M. (2009). Web-based GIS for collaborative planning and public participation: An application to the strategic planning of wind farm sites. *Journal of Environmental Management, 90*(6), 2027–2040. https://doi.org/10.1016/j.jenvman.2007.08.032

St. Martin, K., & Hall-Arber, M. (2008). The missing layer: Geo-technologies, communities, and implications for marine spatial planning. *Marine Policy, 32*(5), 779–786. https://doi.org/10.1016/j.marpol.2008.03.015

Stefanidis, A., Crooks, A., & Radzikowski, J. (2013). Harvesting ambient geospatial information from social media feeds. *GeoJournal, 78*(2), 319–338. https://doi.org/10.1007/s10708-011-9438-2

Stephens, M., & Poorthuis, A. (2015). Follow thy neighbor: Connecting the social and the spatial networks on Twitter. *Computers, Environment and Urban Systems, 53*(September), 87–95. https://doi.org/10.1016/j.compenvurbsys.2014.07.002

Tang, Z., & Liu, T. (2015). Evaluating Internet-based public participation GIS (PPGIS) and volunteered geographic information (VGI) in environmental planning and management. *Journal of Environmental Planning and Management,* (August 2015), 1–18. https://doi.org/10.1080/09640568.2015.1054477

Tulloch, D. L. (2008). Is VGI participation? From vernal pools to video games. *GeoJournal, 72*(3–4), 161–171. https://doi.org/10.1007/s10708-008-9185-1

Turner, A. J. (2006). Introduction to Neogeography. *OReilly Short Cuts*. Retrieved from http://books.google.com/books?hl=en&lr=&id=oHgDv4feV-8C&oi=fnd&pg=PA24&dq=Introduction+to+Neogeography&ots=wYr7RESSeW&sig=HTani7IqwF0NQmiqf8jZigkFguQ

United Nations. (2016). *United Nations E-Government Survey 2016: E-government in support of sustainable development.*

Vasardani, M., Winter, S., & Richter, K.-F. (2013). Locating place names from place descriptions. *International Journal of Geographical Information Science, 27*(12), 2509–2532. https://doi.org/10.1080/13658816.2013.785550

Veljković, N., Bogdanović-Dinić, S., & Stoimenov, L. (2014). Benchmarking open government: An open data perspective. *Government Information Quarterly, 31*(2), 278–290. https://doi.org/10.1016/j.giq.2013.10.011

Weinreb, A. R., & Rofè, Y. (2013). Mapping feeling: An approach to the study of emotional response to built environment and landscape. *Journal of Architectural and Planning Research, 30*(2), 127–145.

Wellman, B. (2002). Little boxes, glocalization, and networked Individualism from little boxes to social networks. *Digital Cities,* 10–25. https://doi.org/10.1007/3-540-45636-8_2

Werts, J. D., Mikhailova, E. a, Post, C. J., & Sharp, J. L. (2012). An integrated WebGIS framework for volunteered geographic information and social media in soil and water conservation. *Environmental Management, 49*(4), 816–32. https://doi.org/10.1007/s00267-012-9818-5

Wiseman, J. (2014). Can 311 call centers improve service delivery? Lessons from New York and Chicago.

Yearley, S., Cinderby, S., Forrester, J., Bailey, P., & Rosen, P. (2003). Participatory modelling and the local governance of the politics of UK air pollution: A three-city case study. *Environmental Values, 12*(2), 247–262. https://doi.org/10.3197/096327103129341315

Zavattaro, S. M., French, P. E., & Mohanty, S. D. (2015). A sentiment analysis of U.S. local government tweets: The connection between tone and citizen involvement. *Government Information Quarterly, 32*(3), 333–341. https://doi.org/10.1016/j.giq.2015.03.003

Zhang, S., & Feick, R. (2016). Understanding public opinions from geosocial media. *ISPRS International Journal of Geo-Information, 5*(6), 74. https://doi.org/10.3390/ijgi5060074

Zhou, X., Xu, C., & Kimmons, B. (2015). Detecting tourism destinations using scalable geospatial analysis based on cloud computing platform. *Computers, Environment and Urban Systems, 54,* 144–153. https://doi.org/10.1016/j.compenvurbsys.2015.07.006

Zook, M. (2017). Crowd-sourcing the smart city: Using big geosocial media metrics in urban governance. *Big Data & Society, 4*(1), 1–13. https://doi.org/10.1177/2053951717694384

Chapter 4
Transactional Participation: Citizen Reporting from Geosocial Media

Abstract Recent developments in volunteered geographic information (VGI) have provided new opportunities for improving city services by enabling citizens instantly and collectively sharing and reporting issues. However, the usefulness of VGI for such use has not been evaluated from a citizen perspective. This chapter aims to bridge this research gap through a case study that innovatively uses geosocial media, as an example of VGI, for reporting accessibility issues to local governments and for providing customized navigation services to the general public. Particularly, a study website was developed that allows citizen participants to evaluate the usefulness of geosocial media for issue reporting and for pedestrian navigation. The results suggest that citizens consider geosocial media useful for helping them maneuver dynamic urban environments and for providing a convenient tool for issue reporting.

4.1 Introduction

Nowadays, local governments are increasingly embarking on ICTs as one way of improving their service delivery and enhancing government–citizen communication and collaboration (Khan, 2015; Wandaogo, 2022; Zhang 2019b;). The rise of Web 2.0 has enabled average citizens to partake in data and content creation and has at least partially transformed the roles of citizens from consumers to co-producers of government data and services (Clark et al., 2013; Sieber & Johnson, 2015). The geospatial web, the geographic component of Web 2.0, provides potential for local governments to augment and improve their services delivery and ultimately to fulfill the goal of smart cities (Cavallo et al., 2014; Karnatak et al., 2022). The geospatial web consists of location-aware applications and services that facilitate the collection and distribution of geospatial information (Masó, 2022). It allows citizens to more directly contribute new types of georeferenced data and to communicate place-based issues. These citizen-generated geospatial data are also considered as volunteered geographic information (VGI), a derivative of user-generated content based on the geospatial web (Elwood et al., 2012).

S. Zhang, *Geo-participation*, https://doi.org/10.1007/978-981-95-8228-0_4

VGI can help urban governors and planners to monitor urban dynamics, to identify deficiencies in public services at finer spatial and temporal resolutions, and to improve the delivery of non-emergency services (Karnatak et al., 2022; Zook, 2017; Zhang 2019b). With location-enabled tools and functions such as geotagging, citizens can conveniently send a request to government for issues like graffiti, parking enforcement, noise complaints, and potholes and include location information. The social networking capabilities of Web 2.0 also enabled local governments to build upon their existing communication channels to foster multidirectional and more dynamic communications among citizens and governments. Taking geosocial media as an example, citizens may collectively report, share and communicate real-time information to both provide feedback to local governments and to assist each other with understanding dynamic urban environments (e.g., road and weather conditions) and co-producing services (Coppola et al., 2022; Linders, 2012; Xu et al., 2022).

Despite this potential, previous studies have identified significant challenges of adopting VGI concerning their uncertain data quality resulted from the lack of supervision in data collection processes, anonymous data authors, inconsistent data formats, and biased representation (Bai et al., 2024; Sieber et al., 2016; Vyron Antoniou, 2023). The uncertain quality of VGI may raise issues including trustworthiness and liability and thus present challenges for local governments to adopt citizen contributions (Johnson, 2016; Khan & Johnson, 2020). However, these conclusions are mostly drawn from government perspectives. As the contributors and users of VGI, how citizens may benefit from collectively generated geospatial information has rarely been studied. Understanding the usefulness of VGI warrants critical evaluation from the perspective of citizen users (MacLean & Titah, 2022; Spicer et al., 2023).

This study intends to bridge this research gap by assessing the usefulness of geosocial media, as a specific example of VGI, for co-producing citizen services and for government–citizen communication. The research objective is addressed by a case study of using geosocial media, for co-producing customized pedestrian navigation services and for citizen reporting sidewalk issues to local governments. Pedestrian infrastructure is an important component of urban infrastructure. Understanding pedestrian dynamics has been an important topic in urban and transportation studies, as it shed light on traffic management and public safety monitoring (Fu et al., 2022; Huo et al., 2022). Access to pedestrian infrastructure is particularly critical for visually- and mobility-impaired people (Fast et al., 2023; Han Qin et al., 2016). Local governments have made considerable efforts including fiscal investment on the regular inspection and maintenance of pedestrian infrastructure such as sidewalks. However, there are still significant challenges for reporting and sharing pedestrian network problems, particularly transient obstacles such as temporary sidewalk block. That is, these temporary problems may make the environment inaccessible but are difficult to capture in a timely manner by routine inspection. Moreover, some groups' (e.g., the visually- and mobility-impaired population) personal needs may not be easily accommodated in general non-emergency reporting programs such as 311, a program for issue reporting and service monitoring by municipal governments in North America (Nam & Pardo, 2014). As such, this chapter examines the potential of using geosocial media for citizens to instantly and collectively share accessibility issues of pedestrian infrastructure. This information can be used both as a report to local governments who are in charge of infrastructure maintenance and for customized navigation services as it is often specific to people with particular mobility conditions.

4.2 Background

4.2.1 Citizens as Co-producers of Government Services

With over 50% of the world's population now living in the cities and the number continuing to grow (United Nations, 2016), cities are increasingly responsible for delivering services to people and businesses (Roche, 2014). Local governments have increasingly capitalized on technologies, particularly Internet and Communication Technologies (ICTs) and Web 2.0 technology, to enhance connections with the public, and improve the efficiency of service delivery by collaborating with the public (Grossman et al., 2018; MacLean & Titah, 2022).

Developments in geospatial technologies, such as the Geoweb and location-aware devices and services, has spurred new forms of government–citizen collaboration. Mirroring the vision of Goodchild (2007) that citizens could act as sensors to revitalize costly expert-driven data systems and to collectively observe a phenomenon and possibly respond faster to the phenomenon, a full spectrum of knowledge producers (e.g., from expert cartographers to lay public) may collectively contribute to government data, services, and infrastructure. For example, citizens may provide up-to-date information (e.g., snow, flooding, traffic) and collaborate with governments on providing customized geolocation services (Linders, 2012; Young, 2022). Inspired by successful crowdsourcing mapping applications, there is a potential for governments to embark on collective mapping practices and involve citizens in updating, correcting, and vetting government geospatial datasets (Bednarska-Olejniczak et al., 2019). According to Sieber and Johnson (2015), this collaboration can be considered a participatory model where data becomes a conduit for integrating citizen contribution into government operations and decision-making. Citizens could also collaborate with governments to produce public services and applications for solving urban problems (Vahidnia, 2022). For instance, citizen collaborate with governments for problem identification and resolution through citizen apps on a variety of topics including transportation, public utilities, health and recreations, and information awareness and access.

4.2.2 VGI for Crowdsourcing Dynamic Urban Environments

VGI has transformed the role of citizens to become creators of spatial data (Goodchild, 2007). Various VGI applications have been developed to harness the collective intelligence of lay people for better capturing dynamic urban environments (Boella et al., 2019). More up-to-date and comprehensive data were collected and provided solutions for disaster response, route planning, infrastructure management, etc. (Goetz & Zipf, 2013; Schade et al., 2013; Vahidnia et al., 2020). Geosocial media, a form of VGI, provides a rich data source for both local governments and citizens to understand dynamic urban environments, as users can collectively and openly generate geospatial information via social media platforms using their innate geotagging functions (König, 2021; Matthews et al., 2023; Wang et al., 2022). A range of studies examined the usefulness of geosocial media in understanding dynamic environments and enhancing situation awareness (Ostermann, 2021; Vahidnia, 2022; Wu, 2020). Other studies have suggested the potential of geosocial media for understanding public perceptions in terms of collective sentiments of local environments and infrastructure (Gugulica & Burghardt, 2023; Xu et al., 2022; Zhang & Feick, 2016). In the context of municipal reporting, geosocial media may reveal problems facing the city that are collectively sensed, detected, and reported by citizens (Wu, 2020). Not only can local governments use this information as a means of collecting citizen feedback, but also can citizens use the information to understand the state of urban environments and assist with their own decision-making (König, 2021; Linders, 2012). For example, a report of a blocked sidewalk is useful both for local governments and for citizens who need to plan the route and avoid the blocked location.

Accessibility issues are often time-critical for citizens to develop situational awareness in a dynamic urban environment. Endeavors have therefore been devoted to use VGI for crowdsourced accessibility mapping. There are a number of VGI applications for crowdsourcing information that may affect pedestrians' accessibility (Fast et al., 2023; Han Qin et al., 2018; Qin et al., 2015). Using the collaboratively collected information, it is now possible to further develop customized pedestrian routing services. Users could share nearly real-time information regarding whether a pedestrian network is accessible or not considering the particular mobility condition of a person. By integrating this information with routing algorithm, people could find the route that best meet their needs.

Considering the potential of geosocial media for understanding dynamic urban environments and for municipal reporting, in this chapter, we explore an innovative use of geosocial media for citizens to collectively share accessibility issues related to pedestrian infrastructure and to report issues to municipal governments. This innovative use will contribute to new forms of government–citizen collaboration by not only producing a service that can help citizens, especially those with mobility challenges, to maneuver dynamic urban environments, but also enabling new forms of citizen reporting that may enhance the responses or awareness of government officials. To explore the potential, the usefulness of geosocial media for such use is examined through a case study in the City of Kitchener, Canada.

4.2.3 Challenges in Adopting VGI

Despite the potential, the uncertain quality of VGI may raise concern for its adoption by local governments (Bai et al., 2024; Khan & Johnson, 2020). Particularly, lack of supervision in data collection processes, anonymous data authors, inconsistent data formats, and biased representation are common challenges identified in previous studies (Basiri et al., 2019). For governments, these quality related challenges raise further concerns regarding the trustworthiness and liability of VGI (Harrison & Johnson, 2019).

Geosocial media information has some common quality issues of VGI such as geographical and thematical unevenness, validity and trustworthiness, and accuracy (Petutschnig et al., 2021). While successful examples have been reported on the usefulness of geosocial media for improving situational awareness of emergency situations (Crooks et al., 2013; Smith et al., 2015), how relevant and useful it is for citizens to understand urban environments and to report non-emergency issues has rarely been investigated. In addition, the aforementioned challenges are mostly identified from local government perspectives. Local government adoption of digital technologies needs to account for various contextual, technological, and organizational challenges (Falco and Kleinhans, 2018). While data quality is one of the major concerns for local government adoption of VGI data, the main challenge associates with how the unassured data quality may lead to issues such as the organizational acceptance of VGI data, the fitness of VGI data into government procedures, etc. (Zhang, 2019a). Citizens' evaluation of VGI data quality may therefore differs from local governments' evaluation, as citizens would focus more on how useful are useful for their needs. It is, however, underexplored how citizens evaluate the usefulness of geosocial media in the context of government–citizen interaction. As such, the main objective of this study is to investigate the innovative use of geosocial media as a means both for citizen co-producing city service and for issue reporting from the perspective of citizens.

4.3 Methods

Citizens' perspectives of the usefulness of geosocial media for crowd sensing dynamic urban environments and for non-emergency reporting were collected through a case study of citizens using geosocial media to share and report on sidewalk conditions. A study website entitled Smart Sidewalks was developed to solicit public opinions toward the usefulness of geosocial media for understanding changing environments and reporting non-emergency issues.

4.3.1 Study Area

The case study was implemented in the downtown of City of Kitchener, Canada (Fig. 4.1). As one of the fastest growing communities in Canada, the City of Kitchener has been endeavoring to pioneer innovative government strategies and improving its service delivery to over 200,000 residents (The City of Kitchener, n.d.).

During the time of the study, the construction of a Light Rail Transit (LRT) in the City of Kitchener added extra complexity to already dynamic sidewalk situations. The two-year construction of the first stage of the project installed 19 km of LRT tracks and 5.6 km of new sidewalks. From December 2016 to March 2017, there were 13 major road closures according to official construction updates, with varying closing times ranging from a day to a few months. However, these updates mainly focused on road closures that were not always the same as sidewalk closures. Pedestrians could hardly know the accessibility of the sidewalk without going to the place. Moreover, proper signage to warn pedestrians with potential barriers and hazards were not always posted in time due to rapidly changing sidewalk conditions (Bueckert, 2016). The signage shown in Fig. 4.2a was reported to be misleading as it directed pedestrians down a blocked street and forced pedestrians to the street (Bueckert, 2016). This brought up serious safety concerns for pedestrians maneuvering construction sites, especially for those with mobility challenges. Situations of temporary pedestrian

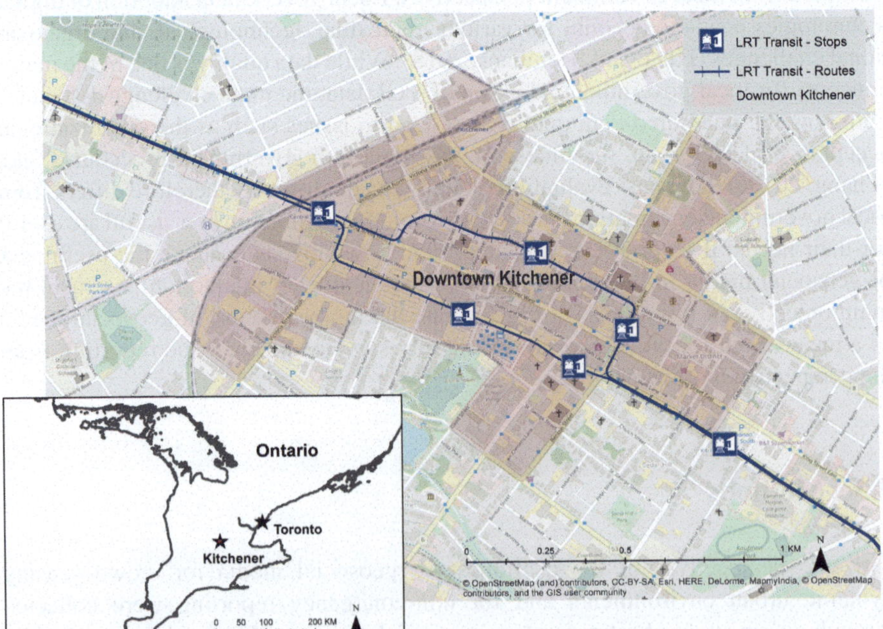

Fig. 4.1 Study area

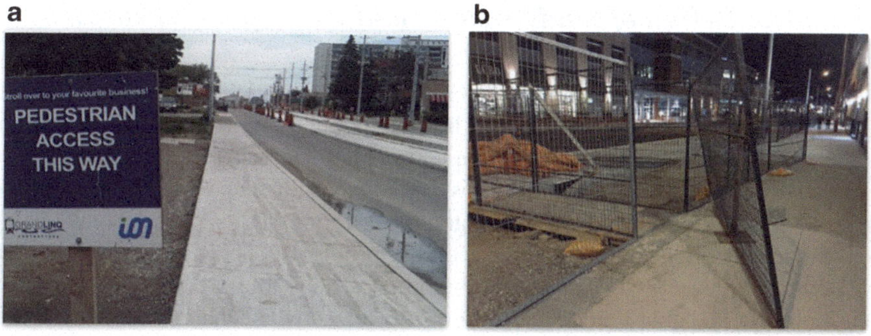

Fig. 4.2 Examples of sidewalk issues in the City of Kitchener: (a) a misleading signage for pedestrian access (*Source* Bueckert, 2016), (b) a temporary pedestrian path that was too narrow for wheelchair users

paths being accessible only to able-bodied pedestrians (Fig. 4.2b) often occurred as well with the construction.

Sidewalk maintenance is an important mandate of local governments. According to minimum maintenance standards in Ontario, sidewalk inspections have to be carried out at least once per year (Ontario, n.d.). In City of Kitchener, Ontario, the municipality spends up to $750,000 for sidewalk inspections and identifies 2800 new sidewalk defects in average every year (City of Kitchener, n.d.). Yet, it is challenging for local governments to capture dynamics of sidewalk conditions by regular inspections. For example, a sidewalk may temporarily be blocked by an obstacle (e.g., construction, poor surface conditions, and vehicles driving on the sidewalk) and be inaccessible for pedestrians. This information is important for both municipalities and citizens important as they may form potential trip obstacles and raise safety risks.

4.3.2 Participant Recruitment

Study participants were recruited during mid-December 2016 to March 2017 through (1) on-street recruitment and (2) contacting municipalities and local organizations. Three on-street recruitments activities were carried out at the City of Kitchener's Farmers Market, one of the major business locations in the study site for distributing project flyers. We also contacted the City of Kitchener and local community organizations working on accessibility related issues to help broadcast project information through flyers and social media. The study was open to participants with a minimum age of 14 years old without restrictions on their own locale. However, given that the public involvement is local in nature (Kidney, 2002), participants were mostly local residents.

4.3.3 The Design of Study Website and Participation Process

The study website consists of two parts. First, Twitter and Flickr are used as examples of geosocial media considering that they both are popular platforms but have different focuses (i.e., Twitter is a timely communication platform, whereas Flickr focuses on photo-sharing). The messages and photos displayed on the map were harvested based on keywords (LRT, sidewalk) and geographic locations (City of Kitchener) using the Twitter and Flickr public APIs. Second, a proof-of-concept pedestrian navigation service is developed. The service help pedestrians find proper routes that suit their mobility (i.e., able-bodied, cane users, motorized wheelchair users, and manual wheelchair users) by considering important factors, including elevation, sidewalk width, surface materials, that affect pedestrians. Users can further add perceived barriers to routing choices if they identify one or more barriers by examining geosocial media messages. The service would then find users an alternative route that avoid perceived barriers.

Participants were asked to access the study website Smart Sidewalks at their own time and location. Participants would follow the step-by-step guidance on the website to understand how they may use geosocial media for sharing and reporting sidewalk conditions and complete the survey accordingly. First, participants were asked to examine at least five Twitter messages and five Flickr photos within a selected construction area of interest on the map and determined how relevant these message and photos to sidewalk conditions (Fig. 4.3a). Second, participants used the pedestrian routing service to examine if and how geosocial media is useful for them to maneuver around construction sites (Fig. 4.3b and c).

Previous studies have suggested the key dimensions of VGI data quality, including the semantic accuracy, temporal accuracy, logical consistency, completeness (Antoniou & Skopeliti, 2015). Users' characteristic such as their credibility is also an important indicator of VGI data quality, particularly with regard to social media platform (Banerjee et al., 2017; Hung et al., 2016). As such, we design a questionnaire asking participants about the usefulness of geosocial media for understanding dynamic urban environment regarding its semantic accuracy, relevance, locational accuracy, trustworthiness, and timeliness. The survey also asked participants general questions regarding their experiences with citizen reporting and their perceived opportunities and challenges of geosocial media reporting. Participants were asked to complete the online survey held on SurveyMonkey once they finished using the Smart Sidewalks application.

4.3.4 Data Analysis

The analysis focuses on the usefulness of geosocial media for pedestrian navigation and for non-emergency reporting respectively. For the former, the evaluation focuses on five spatial data quality aspects including relevance, semantic accuracy, locational

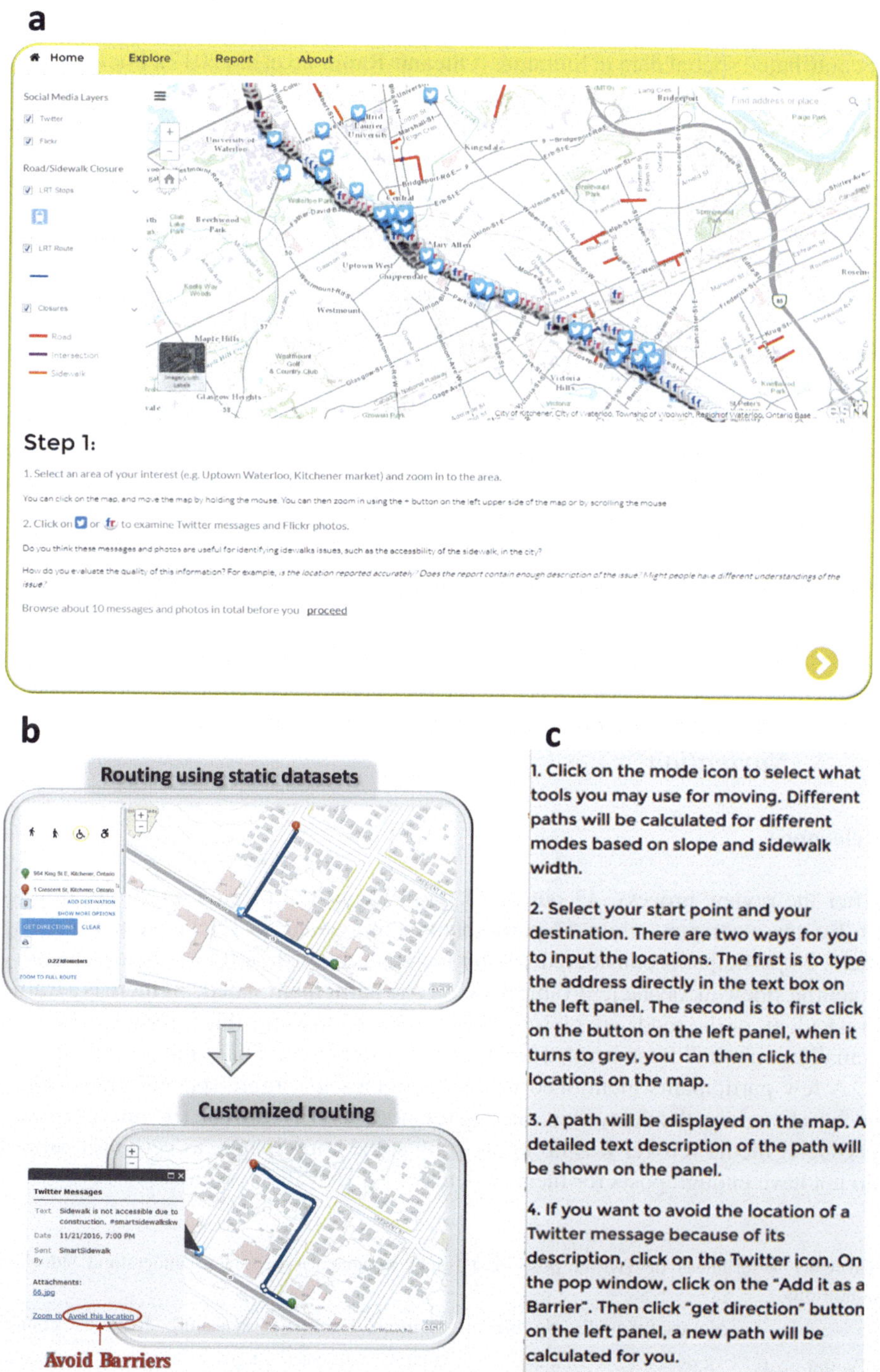

Fig. 4.3 Study website: (a) Map display of Twitter messages and Flickr photos, (b) Proof-of-concept service: Social media messages as aids to plan pedestrian routes, (c) Instruction for participants to use the pedestrian service

accuracy, trustworthiness, and timeliness, as they are most addressed quality aspects of contributed spatial data in literature (Olteanu-Raimond et al., 2017). For the latter, opportunities and challenges of geosocial media reporting were investigated based on the percentage of the total number of participants that agree or disagree with each given statement regarding the usefulness of geosocial media reporting.

4.4 Results

4.4.1 Study Sample

A total of 47 participants completed the assigned tasks and survey questions. Nearly half of the participants (45.9%) fell into the 25–35 years' age group, and 20.5% of participants were between 35 and 45. The other age groups have very similar shares of total participants, with 11.5% of participants under 25, 9.1% aged between 45 and 55, and 11.3% over 55. The distribution of male and female participants was almost equal (44.4% and 48.9% respectively, the rest 6.7% participants would not like to disclose their gender information).

4.4.2 Usefulness of Geosocial Media for Pedestrian Navigation

Relevance

After the review process, 45 out of 47 participants commented on the usefulness of Twitter messages, while 42 participants commented on Flickr photos. Among them, 28 participants considered Twitter messages being helpful for them to identify potential sidewalk obstacles (Table 4.1). As one participant stated, "If there is a traffic or accident, or the roads are closed, then it is useful to know this information before actually going in." In contrast, 12 participants considered Flickr photos useful.

A few participants mentioned in the comments that more specific information, such as how long the delay is, is needed for sidewalk navigation. The unevenness of geosocial media reports was brought up by some participants, as certain locations do not have enough posts for them to determine sidewalk conditions.

Table 4.1 Usefulness of Twitter posts and Flickr images for people to understand sidewalk conditions

	Not useful at all	Not useful	Neutral	Useful	Very useful	Response count
Twitter posts	8.9%	4.4%	24.4%	48.9%	13.3%	45
Flickr images	16.7%	7.1%	47.6%	26.2%	2.4%	42

Semantic Accuracy

Participants raised the issue of the subjectivity of geosocial media. As one participant said in the comments, geosocial media might be too prone to subjective commentary of text messages, or selective composition of photos. One of our retrieved tweets said that "intersection at (street A and B) is not accessible for wheelchair users." It is yet unknown whether this intersection is accessible for able-bodied people or cane users.

Twenty-six participants indicated that they are more likely to consider a post that has photo attachments to be useful. Five participants suggested in comments that pictures are more useful for them to understand real-world circumstances, whereas text descriptions are sometimes less informative or less accurate.

Locational Accuracy

The location information associated with social media posts was considered a valuable asset. Twenty-eight participants believed that the geotagging function allows accurate communication of the locations of sidewalk problems. However, answers to the question "What challenges do you find when adding location information to social media posts?" suggested that nearly half (48.8%) of the participants do not want to share their location information. This concern with privacy is not new (Elwood & Leszczynski, 2011). Previous studies have similarly suggested that smartphone users either did not know about geotagging function or would not like to disclose their location information (Ricker et al., 2015). Accordingly, insufficient numbers of people sharing geolocated information about sidewalks can be a challenge for using geosocial media as a dynamic source of understanding urban environment. Moreover, half of the ten participants aged over 45 claimed that they do not know how to add location information to social media. While social media is often considered easy to use, our results suggest that certain functions of social media such as geotagging may not be well accepted by all of the population groups.

Trustworthiness

The issue of reliability of geosocial media was raised by participants; 72.3% of the participants ($n = 34$) claimed that they are more likely to trust the posts coming from official accounts (e.g., local government). In the comments, some participants stated that they regularly check @570 Traffic, a Twitter account operated by a local radio station that updates local road conditions on an hourly basis. Participants suggested that a similar method of assembling Twitter messages would increase the reliability of the information as well as reduce the effort of looking for specific information.

Timeliness

According to 75% of the participants ($n = 32$), timeliness is an important indicator for them to evaluate the usefulness of geosocial media reports. Not being timely is a major factor that makes Flickr less useful, as it is a photo-sharing website where users do not always post at the time and location of a photo is taken whereas Twitter is more used for immediate communication.

4.4.3 Usefulness of Geosocial Media for Non-emergency Reporting

With regard to geosocial media reporting, participants had very positive attitude, considering geosocial media provides an easy form of reporting and can stimulate effective and open communication (Table 4.2).

Opportunities of Geosocial Media Reporting

Not surprisingly, participants considered that reporting via social media might be easier than reporting via phone calls or reporting applications. 84.8% and 78.3% of the participants believed that social media allows them to report non-emergency issues at any time and location respectively. Participants also believed that they can easily add pictures and location information to help city staff identify what and where a reported issue is; 63% of participants considered direct contact with city being convenient for them to report issues, as they do not need to look for specific reporting methods (e.g., phone numbers, website). When asked about whether they have reported issues to the city before in a separate question, 47.4% of the participants claimed that they have never reported non-emergency issues. Among them, 54.2% indicated that reporting was too much effort for them. The ease of social media reporting thus can potentially mobilize these people to increase reporting rates.

Among those participants who have reported to the city before, 23.2% claimed that they were not satisfied or extremely not satisfied with their reporting experience. Slow response from governments, in particular, was the most mentioned reason that dissatisfied people. In contrast, citizens anticipated social media reporting to be more efficient. A majority of the participants (73.9%) believed that being able to receive up-to-date reports from citizens allows governments to react more promptly. In addition, 65.2% of the participants considered the openness of social media as advantageous for expanding government–citizen communication to citizen-citizen information sharing. As suggested by Linders (2012), this greater interactivity may enhance citizen engagement and foster innovative forms of collaboration.

Table 4.2 Percentages of participants that consider the following statements to be the advantages or challenges of using geosocial media reporting

	Theme	Statement	Response percentage
Opportunities	An easy form of reporting	Citizens can report issues at any time	84.8%
		Citizens can report issues from anywhere	78.3%
		It is easy to add pictures to social media to help city staff to understand the problem	80.4%
		It is easy to add exact locations to social media so that citizens can report the location of issues easily and precisely	63.0%
		The cost of reporting via social media is low	69.6%
		Citizens do not need to find contact information of the city to report	63.0%
	Effective and open communication	City staff can get most up-to-date reports and thus may respond more efficiently	73.9%
		Many people are using social media so that city staff can get more feedback from citizens	60.9%
		Social media is an open communication platform so that citizens can see each other's reports as well as city staff's response	65.2%
Challenges	Official acceptance	It is not an official reporting method, and government officials may not respond	42.2%
		It is not an official reporting method, and government officials may not treat these reports the same way as reports coming from other methods	44.4%
		The city will need to devote more personnel to processing and monitoring social media and is less likely to consider social media as a reporting method	40.0%
		I do not know how to report via social media (e.g., use what format, keywords, hashtags)	15.6%
	Quality issues	People often post emotional messages rather than describe what issues are on social media, therefore the reports may not be as trustworthy as reports coming from other methods	42.2%
		Sometimes the situation is too complicated to describe within a word limit (e.g., each Twitter message can only have 140 words maximum)	35.6%

(continued)

Table 4.2 (continued)

	Theme	Statement	Response percentage
		Citizens' personal smartphone GPS may not be sufficiently precise, thus locations reported to city staff may not be accurate	22.2%
	Privacy	I do not want to make my report viewable to the public	15.6%
		I don't want to add location information, because I don't want other people know where I am	17.2%
	Others	I am more used to other reporting methods (e.g., phone, email) and would like to continue using them	11.1%

Challenges of Geosocial Media Reporting

Comparatively, a smaller portion of participants (less than 50%) agreed with statements regarding challenges of using geosocial media reporting. The major concern participants had with geosocial media reporting is that it is not an official reporting method, and that government officials may not respond to geosocial media reports or not respond similarly to they would do with reports from other channels. Along a similar line, 42.2% of the participants agreed that the often too emotional tone of social media messages could make geosocial media reporting less trustworthy compared to other reporting methods. Other participants indicated that the details and locations of non-emergency issues may not be communicated clearly and precisely through social media.

Answers to privacy concerns revealed some interesting results. Only 15.6% and 17.2% of the participants claimed that they would not like to make their report viewable to public or disclose their location information, respectively. These results support citizen participants' beliefs that geosocial media provide an open environment for efficient communication. However, it contrasts to the answers regarding challenges of using geosocial media, for which nearly half of the participants claimed that they do not want to share their locations. These seemingly contradictory results have some indications that the understanding of privacy is context-dependent. While people may not want to share personal location information such as their home addresses, they may not have issues with adding location information when reporting an issue at public places.

While only five participants state that they would prefer to continue using reporting methods that they have used before, 60% of them were in the age group of above 55. Consistent with other studies about digital divide (Lněnička & Máchová, 2022), this result can shed light on the possibility that different population groups may accept geosocial media reporting differently.

4.5 Discussion

4.5.1 Challenges of Using VGI

Our study suggests some common challenges of using VGI. One is the inherent bias of VGI (Petutschnig et al., 2021). In our study, despite the extensive recruitment efforts, the sample size of survey participants is small. The small sample is limited in representing the broader population, yet it exhibits the often-skewed representation of participants as found in other VGI studies (Haworth et al., 2016). In addition, the uneven spatial distribution of geosocial media posts harvested from Twitter and Flickr indicates the selective representation of geosocial data as suggested in other studies (Zook, 2017). Crowdsourced data is almost by default related to personal interests and experiences (Mooney et al., 2011). In our study, half of the participants are between 25 and 35; people aged over 55 showed less likelihood of using geotagging function because of privacy concerns and limited experiences with the technology. As a result, geosocial media information may only show interests of certain population groups and are limited by geographic locations that are valued by individuals who are more frequent users of social media.

The second is the privacy issue, a major concern that may prevent citizens from contributing data (Zook, 2017). In addition to commonly expressed concern of disclosing personal trajectories, our study has implications that the understanding of privacy is context-specific. People are more willing to provide geolocation information when they know the information is used for sharing and reporting non-emergency issues. Lacking knowledge about privacy was identified as a critical factor that prevents people from using online participatory services (Christin et al., 2011). In addition, the public is more willing to use the technology when perceived benefits outweigh potential risks (Ricker et al., 2015). In line with these comments, we may develop strategies of alleviating public's privacy concerns, such as providing participants with details of benefits and risks associated with sharing location information and develop location-based services that are directed to specific uses, to encourage participation.

4.5.2 Using VGI for Customized Citizen Services and Issue Reporting

Despite the challenges, this study indicates the opportunities of using geosocial media, as a case of VGI, for developing customized citizen services. From the geospatial data perspective, citizens find geosocial media helpful for wayfinding albeit a few deficiencies. This suggests the potential of developing an immediate and collaborative form of civic action. That is, citizens can determine the usefulness of information based on their own criteria and needs. As shown in the case study, data contributed by one person may only be of value to a group of users that share similar interests or have similar needs. As such, personalized information that is tailored to individual needs could be delivered through citizen collaboration. To some extent, this collective information sharing may extend government services as citizens can act upon the information without the discretion of governments and generate their own services outside of government mandate (Linders, 2012).

With regard to issue reporting, citizens would like to adopt geosocial media reporting as it is a convenient form of reporting and it enables open and direct communication among multiple stakeholders. However, it is unknown whether or not geosocial media reporting is acknowledged by local governments. Citizens are concerned that the report will not be properly handled if the method is not officially accepted. In fact, local governments face significant quality and credibility related challenges of adopting citizen contributed data (Harrison & Johnson, 2019; Johnson et al., 2017). The contradiction indicates that different users (i.e., citizens and governments) may have diverging perspectives toward the quality issues of VGI and accordingly its usefulness. To better take advantage of VGI for customized citizen services, the incorporation of data validation or quality control measures are necessary. In fact, data quality has been a central focus of VGI studies (Senaratne et al., 2017). Various data quality assessment methods have been proposed, accounting for various factors including data accuracy (e.g., semantic, locational, temporal and logical accuracy), socioeconomic characteristics, and contributors' characteristics (e.g., users' trustworthiness and their data contribution behavior) (Antoniou & Skopeliti, 2015). More recently, developments in data-centric artificial intelligence (DCAI) technology provide new opportunities for improving the quality of geospatial data through AI-enabled data labeling and cross-validation (Wu et al., 2024). Future work may consider developing hybrid data quality assessment and validation measures that can automatically evaluate the accuracy of data and optimize data quality, so that concerns regarding the uncertain quality of VGI can partially be mitigated.

4.6 Conclusions

This chapter investigates the potential of geosocial media, as a form of VGI, for developing customized citizen services and for non-emergency reporting from the perspective of citizens. The case study of citizen crowdsourcing sidewalk issues suggests that citizens consider geosocial media to be a valuable source for understanding dynamic urban environments. This finding deviates from the common critique regarding the uncertain quality of VGI data and indicates the potential of using VGI for developing novel citizen collaborations. In addition, the study suggests that geosocial media may provide an open and convenient conduit for citizens to communicate non-emergency issues with local governments. While organizational and ethical factors need to be further considered based on local governments' perspectives, the preference of the public should be taken into account as local governments aim to improve their service response and citizen satisfaction. Overall, this study contributes to the broad discussions of citizen co-producing services in the context of evolving government–citizen relationships. Different from previous studies that often draw upon local government perspective, the investigation of citizen perspective suggests that some common barriers, particularly the quality issues and privacy concerns, to the use of citizen contributed data are context-specific. This nuanced understanding of VGI is necessary for the proper use of citizen collaborations.

Despite the contributions, we have to acknowledge the limitation that the study was implemented in 2017, when the ICT use in the context of e-government and smart cities might have greatly changed. However, the issue discussed in this study still pertain in the current smart city developments, particularly considering the citizen-centric focus of smart city. In addition, Canadian governments were among the earliest in initiating government–citizen collaboration through citizen co-producing data and services. The lessons learnt from this study may shed light on the broad citizen-centric smart city initiatives. We also envision several directions that worth further investigations based on our research findings. First, the study sample of citizen participants tends to over-represent people who are aged between 25 and 35. Other studies have characterized the majority of social media users as young, well-educated and more affluent and suggested the potential of social media as channel of engaging young population who are often under-represented in participatory projects (Schweitzer, 2014). It is necessary to further investigate the divisions in how people use geosocial media, particularly concerning the vulnerable groups (e.g., the elderly and people with visual impairment) as they may encounter challenges using the technology. To outreach to these under-represented groups, VGI applications may also include welfare technology such as audio–video contact and impairments compensation, to provide communication support and assistance (Iancu & Iancu, 2020). Second, our study suggests the potential of citizen developing their own services considering the differences between how governments and citizens evaluate and use information. To further explore this potential, we may also consider how citizens will use information differently. For instance, how skewed representation of users may affect the delivery of personalized service needs to be further unraveled in future studies. Third, although 47 participants represent a reasonable sample for public participation studies, future research may further expand the sample size to include more diverse population groups, especially the vulnerable groups, and those from different age group and geographic areas, so that a more comprehensive understanding with regard to implications for developing citizen-centric services can be obtained.

References

Antoniou, V., & Skopeliti, A. (2015). Measures and indicators of VGI quality: An overview. *ISPRS Annals of the Photogrammetry, Remote Sensing and Spatial Information Sciences, II-3-W5*, 345–351. https://doi.org/10.5194/isprsannals-II-3-W5-345-2015.

Antoniou, V. (2023). On volunteered geographic information quality: A framework for sharing data quality information. In *Geoinformatics for geosciences* (pp. 149–160). Elsevier. https://doi.org/10.1016/B978-0-323-98983-1.00009-0.

Bai, A., Satarpour, M., Mohebbi, F., & Forati, A. M. (2024). Digital crowdsourcing and VGI: Impact on information quality and business intelligence. *Spatial Information Research, 32*(4), 463–471. https://doi.org/10.1007/s41324-024-00572-2

Banerjee, S., Bhattacharyya, S., & Bose, I. (2017). Whose online reviews to trust? Understanding reviewer trustworthiness and its impact on business. *Decision Support Systems, 96,* 17–26. https://doi.org/10.1016/j.dss.2017.01.006

Basiri, A., Haklay, M., Foody, G., & Mooney, P. (2019). Crowdsourced geospatial data quality: Challenges and future directions. *International Journal of Geographical Information Science, 33*(8), 1588–1593. https://www.tandfonline.com/doi/abs/10.1080/13658816.2019.1593422.

Bednarska-Olejniczak, D., Olejniczak, J., & Svobodová, L. (2019). Towards a smart and sustainable city with the involvement of public participation—The case of Wroclaw. *Sustainability, 11*(2), 332. https://doi.org/10.3390/su11020332

Boella, G., Calafiore, A., Grassi, E., Rapp, A., Sanasi, L., & Schifanella, C. (2019). Firstlife: Combining social networking and VGI to create an urban coordination and collaboration platform. *IEEE Access, 7,* 63230–63246. https://doi.org/10.1109/ACCESS.2019.2916578

Bueckert, K. (2016, June 19). *Pedestrian safety near LRT construction sites galls residents.* CBC News. http://www.cbc.ca/news/canada/kitchener-waterloo/kitchener-waterloo-lrt-construction-sites-pedestrian-safety-concern-1.3640705

Cavallo, S., Lynch, J., & Scull, P. (2014). The digital divide in citizen-initiated government contacts: A GIS approach. *Journal of Urban Technology, 21*(4), 77–93. https://doi.org/10.1080/10630732.2014.942167

Christin, D., Reinhardt, A., Kanhere, S. S., & Hollick, M. (2011). A survey on privacy in mobile participatory sensing applications. *Journal of Systems and Software, 84*(11), 1928–1946. https://doi.org/10.1016/j.jss.2011.06.073

City of Kitchener. (n.d.). *Roads and sidewalks.* Retrieved August 15, 2017, from. https://www.kitchener.ca/en/livinginkitchener/RoadsAndSidewalks.asp.

Clark, B. Y., Brudney, J. L., Jakobsen, M., & Andersen, S. C. (2013). Coproduction of government services and the new information technology: investigating the distributional biases. *Public Administration Review, 73*(5), 704–713. https://doi.org/10.1111/puar.12092.Coproduction

Coppola, F., Grimaldi, M., & Fasolino, I. (2022). Constructing dynamic scenarios of crime risk exposure: A methodological proposal based on geo-social media data. In E. Borgogno-Mondino & P. Zamperlin (Eds.), *Geomatics for green and digital transition* (pp. 156–165). Springer. https://doi.org/10.1007/978-3-031-17439-1_11.

Crooks, A., Croitoru, A., Stefanidis, A., & Radzikowski, J. (2013). #Earthquake: Twitter as a distributed sensor system. *Transactions in GIS, 17*(1), 124–147. https://doi.org/10.1111/j.1467-9671.2012.01359.x

Elwood, S., & Leszczynski, A. (2011). Privacy, reconsidered: New representations, data practices, and the GeoWeb. *Geoforum, 42*(1), 6–15. https://doi.org/10.1016/j.geoforum.2010.08.003

Elwood, S., Goodchild, M. F., & Sui, D. Z. (2012). Researching volunteered geographic information: Spatial data, geographic research, and new social practice. *Annals of the Association of American Geographers, 102*(3), 571–590. https://doi.org/10.1080/00045608.2011.595657

Falco, E., & Kleinhans, R. (2018). Beyond technology: Identifying local government challenges for using digital platforms for citizen engagement. *International Journal of Information Management, 40,* 17–20. https://doi.org/10.1016/j.ijinfomgt.2018.01.007.

Fast, V., Lamoureux, Z., & Derksen, L. (2023). Mapping for access: A methodology for improving inclusion on university campuses. *Cartographica: The International Journal for Geographic Information and Geovisualization, 58*(1), 1–20. https://doi.org/10.3138/cart-2021-0021.

Fu, X., Yu., Guanyi, & Liu, Z. (2022). Spatial–temporal convolutional model for urban crowd density prediction based on mobile-phone signaling data. *IEEE Transactions on Intelligent Transportation Systems, 23*(9), 14661–14673. https://doi.org/10.1109/TITS.2021.3131337

Goetz, M., & Zipf, A. (2013). Indoor route planning with volunteered geographic information on a (mobile) web-based platform. In J. M. Krisp (Ed.), *Progress in location-based services* (pp. 211–231). Springer. https://doi.org/10.1007/978-3-642-34203-5_12.

Goodchild, M. F. (2007). Citizens as sensors: The world of volunteered geography. *GeoJournal, 69*(4), 211–221. https://doi.org/10.1007/s10708-007-9111-y

Grossman, G., Platas, M. R., & Rodden, J. (2018). Crowdsourcing accountability: ICT for service delivery. *World Development, 112*(December), 74–87. https://doi.org/10.1016/j.worlddev.2018.07.001

Gugulica, M., & Burghardt, D. (2023). Mapping indicators of cultural ecosystem services use in urban green spaces based on text classification of geosocial media data. *Ecosystem Services, 60,* Article 101508. https://doi.org/10.1016/j.ecoser.2022.101508.

Harrison, S., & Johnson, P. (2019). Challenges in the adoption of crisis crowdsourcing and social media in Canadian Emergency Management. *Government Information Quarterly, 36*(3), 501–509. https://doi.org/10.1016/j.giq.2019.04.002

Haworth, B., Whittaker, J., & Bruce, E. (2016). Assessing the application and value of participatory mapping for community bushfire preparation. *Applied Geography, 76,* 115–127. https://doi.org/10.1016/j.apgeog.2016.09.019

Hung, K.-C., Kalantari, M., & Rajabifard, A. (2016). Methods for assessing the credibility of volunteered geographic information in flood response: A case study in Brisbane, Australia. *Applied Geography, 68,* 37–47. https://doi.org/10.1016/j.apgeog.2016.01.005.

Huo, J., Fu., Xiao, Liu, Z., Qi., & Zhang. (2022). Short-term estimation and prediction of pedestrian density in urban hot spots based on mobile phone data. *IEEE Transactions on Intelligent Transportation Systems, 23*(8), 10827–10838. https://doi.org/10.1109/TITS.2021.3096274

Iancu, I., & Iancu, B. (2020). Designing mobile technology for elderly. A theoretical overview. *Technological Forecasting and Social Change, 155*(June), Article 119977. https://doi.org/10.1016/j.techfore.2020.119977

Johnson, P. A. (2016). Models of direct editing of government spatial data: Challenges and constraints to the acceptance of contributed data. *Cartography and Geographic Information Science, 0406*(May), 1–11. https://doi.org/10.1080/15230406.2016.1176536

Johnson, P. A., Sieber, R. E., Scassa, T., Stephens, M., & Robinson, P. (2017). The cost(s) of geospatial open data. *Transactions in GIS, 21*(3), 434–445. https://doi.org/10.1111/tgis.12283

Karnatak, H., Pandey, K., & Raghavaswamy, V. (2022). Online geodata repositories, geoweb services and emerging geospatial technologies for smart city planning. In R. K. Mishra, C. L. Kumari, S. Chachra, P. S. J. Krishna, A. Dubey, & R. B. Singh (Eds.), *Smart cities for sustainable development* (pp. 211–229). Springer. https://doi.org/10.1007/978-981-16-7410-5_13.

Khan, G. F. (2015). The government 2.0 utilization model and implementation scenarios. *Information Development, 31*(2), 135–149. https://doi.org/10.1177/0266666913502061

Khan, Z. T., & Johnson, P. A. (2020). Citizen and government co-production of data: analyzing the challenges to government adoption of VGI—Khan—2020—Canadian geographies/géographies Canadiennes—Wiley Online Library. *The Canadian Geographer/le Géographe Canadiener, 64*(3), 374–387.

Kidney, T. C. (2002). *Public involvement and civic rationalism in local authority planning and decision making.* University of Wales.

König, P. D. (2021). Citizen-centered data governance in the smart city: From ethics to accountability. *Sustainable Cities and Society, 75*(December), Article 103308. https://doi.org/10.1016/j.scs.2021.103308

Linders, D. (2012). From e-government to we-government: Defining a typology for citizen coproduction in the age of social media. *Government Information Quarterly, 29*(4), 446–454. https://doi.org/10.1016/j.giq.2012.06.003

Lněnička, M., & Máchová, R. (2022). A theoretical framework to evaluate ICT disparities and digital divides: Challenges and implications for e-government development. *Review of Economic Perspectives, 22*(1), 25–51. https://sciendo.com/article/10.2478/revecp-2022-0002.

MacLean, D., & Titah, R. (2022). A systematic literature review of empirical research on the impacts of E-government: A public value perspective. *Public Administration Review, 82*(1), 23–38. https://doi.org/10.1111/puar.13413

Masó, J. (2022). Geospatial web services. In W. Kresse & D. Danko (Eds.), *Springer handbook of geographic information* (pp. 493–530). Springer. https://doi.org/10.1007/978-3-030-53125-6_16.

Matthews, P., Parsons, A., Nyanzu, E., & Rae, A. (2023). Dog fouling and potholes: understanding the role of coproducing 'citizen sensors' in local governance. *Local Government Studies, 49*(5), 908–931. https://doi.org/10.1080/03003930.2022.2116575

Mooney, P., Sun, H., & Yan, L. (2011). VGI as a dynamically updating data source in location-based services in urban environments. In *Proceedings of the 2nd International Workshop on Ubiquitous Crowdsourcing* (pp. 1027–1036). ACM. https://doi.org/10.1145/1357054.1357213.

Nam, T., & Pardo, T. A. (2014). The changing face of a city government: A case study of Philly311. *Government Information Quarterly, 31*(Suppl. 1), S1–S9. https://doi.org/10.1016/j.giq.2014.01.002.

Olteanu-Raimond, A.-M., Hart, G., Foody, G. M., Touya, G., Kellenberger, T., & Demetriou, D. (2017). The scale of VGI in map production: A perspective on European national mapping agencies. *Transactions in GIS, 21,* 74–90. https://doi.org/10.1111/tgis.12189

Ontario. (n.d.). "ONTARIO REGULATION 239/02: MINIMUM MAINTENANCE STANDARDS FOR MUNICIPAL HIGHWAYS."

Ostermann, F. O. (2021). Linking geosocial sensing with the socio-demographic fabric of smart cities. *ISPRS International Journal of Geo-Information, 10*(2), 52. https://doi.org/10.3390/ijgi10020052

Petutschnig, A., Resch, B., Lang, S., & Havas, C. (2021). Evaluating the representativeness of socio-demographic variables over time for geo-social media data. *ISPRS International Journal of Geo-Information, 10*(5), 323. https://doi.org/10.3390/ijgi10050323

Qin, H., Aburizaiza, A. O., Rice, R. M., Paez, F., & Rice, M. T. (2015). Obstacle characterization in a geocrowdsourced accessibility system. *ISPRS Annals of the Photogrammetry, Remote Sensing and Spatial Information Sciences, II-3-W5,* 179–185. https://doi.org/10.5194/isprsannals-II-3-W5-179-2015.

Qin, H., Rice, R. M., Fuhrmann, S., Rice, M. T., Curtin, K. M., & Ong, E. (2016). Geocrowdsourcing and accessibility for dynamic environments. *GeoJournal, 81*(5), 699–716. https://doi.org/10.1007/s10708-015-9659-x

Qin, H., Curtin, K. M., & Rice, M. T. (2018). Pedestrian network repair with spatial optimization models and geocrowdsourced data. *GeoJournal, 83*(2), 347–364. https://doi.org/10.1007/s10708-017-9775-x

Ricker, B., Schuurman, N., & Kessler, F. (2015). Implications of smartphone usage on privacy and spatial cognition: Academic literature and public perceptions. *GeoJournal, 80*(5), 637–652. https://doi.org/10.1007/s10708-014-9568-4

Roche, S. (2014). Geographic information science I: Why does a smart city need to be spatially enabled? *Progress in Human Geography, 38*(5), 703–711.

Schade, S., Díaz, L., Ostermann, F., Spinsanti, L., Luraschi, G., Cox, S., Nuñez, M., & De Longueville, B. (2013). Citizen-based sensing of crisis events: Sensor Web Enablement for volunteered geographic information. *Applied Geomatics, 5*(1), 3–18. https://doi.org/10.1007/s12518-011-0056-y

Schweitzer, L. (2014). Planning and social media: A case study of public transit and stigma on Twitter. *Journal of the American Planning Association, 80*(3), 218–238. https://doi.org/10.1080/01944363.2014.980439

Senaratne, H., Mobasheri, A., Ali, A. L., Capineri, C., & Haklay, M. (2017). A review of volunteered geographic information quality assessment methods. *International Journal of Geographical Information Science, 31*(1), 139–167. https://doi.org/10.1080/13658816.2016.1189556

Sieber, R. E., & Johnson, P. A. (2015). Civic open data at a crossroads: Dominant models and current challenges. *Government Information Quarterly, 32*(3), 308–315. https://doi.org/10.1016/j.giq.2015.05.003

Sieber, R. E., Robinson, P. J., Johnson, P. A., & Corbett, J. M. (2016). Doing public participation on the Geospatial Web. *Annals of the American Association of Geographers, 106*(5), 1030–1046. https://doi.org/10.1080/24694452.2016.1191325

Smith, L., Liang, Q., James, P., & Lin, W. (2015). Assessing the utility of social media as a data source for flood risk management using a real-time modelling framework. *Journal of Flood Risk Management,* n/a-n/a. https://doi.org/10.1111/jfr3.12154

Spicer, Z., Goodman, N., & Wolfe, D. A. (2023). How 'smart' are smart cities? Resident attitudes towards smart city design. *Cities, 141*(October), Article 104442. https://doi.org/10.1016/j.cities.2023.104442

The City of Kitchener. (n.d.). *Kitchener's Strategic Plan 2015–2018.*

United Nations. 2016. *United Nations e-government survey 2016: E-government in support of sustainable development.*

Vahidnia, M. H. (2022). Citizen participation through volunteered Geo-Graphic information as equipment for a smart city to monitor urban decay. *Environmental Monitoring and Assessment, 195*(1), 181. https://doi.org/10.1007/s10661-022-10796-0

Vahidnia, M. H., Hosseinali, F., & Shafiei, M. (2020). Crowdsource mapping of target buildings in hazard: The utilization of smartphone technologies and geographic services. *Applied Geomatics, 12*(1), 3–14. https://doi.org/10.1007/s12518-019-00280-9

Wandaogo, A.-A. (2022). Does digitalization improve government effectiveness? Evidence from developing and developed countries. *Applied Economics, 54*(33), 3840–3860. https://www.tandfonline.com/doi/abs/10.1080/00036846.2021.2016590.

Wang, Y., Nagireddy, S. R., Thota, C. T., Ho, D. H., & Lee, Y. (2022). Community-in-the-loop: Creating artificial process intelligence for co-production of city service. *Proceedings of the ACM on Human-Computer Interaction, 6*(CSCW2), 1–21. https://doi.org/10.1145/3555176

Wu, H., Jiang, Z., Dong, A., Gao, R., Yan, X., Hu, Z., Mao, F., & Lin, Y. (2024). DCAI-CLUD: A data-centric framework for the construction of land-use datasets. *International Journal of Geographical Information Science, 38*(11), 2205–2232. https://www.tandfonline.com/doi/abs/10.1080/13658816.2024.2387200.

Wu, W.-N. (2020). Determinants of citizen-generated data in a smart city: Analysis of 311 system user behavior. *Sustainable Cities and Society, 59*(August), Article 102167. https://doi.org/10.1016/j.scs.2020.102167

Xu, S., Li, S., Huang, W., & Wen, R. (2022). Detecting spatiotemporal traffic events using geosocial media data. *Computers, Environment and Urban Systems, 94,* Article 101797. https://doi.org/10.1016/j.compenvurbsys.2022.101797.

Young, M. M. (2022). The impact of technological innovation on service delivery: Social media and smartphone integration in a 311 system. *Public Management Review, 24*(6), 926–950. https://doi.org/10.1080/14719037.2021.1877794

Zhang, S. (2019). Public participation in the GeoWeb era: Defining a typology for Geo- participation in local govern-ments. *Cities, 85,* 38–50. https://doi.org/10.1016/j.cities.2018.12.004

———. 2019b. Public participation in the geoweb era: Geosocial media use in local government. *Cities, 85*(Feburary), 38–50. https://doi.org/10.1016/j.cities.2018.12.004.

Zhang, S., & Feick, R. (2016). Understanding public opinions from geosocial media. *ISPRS International Journal of Geo-Information, 5*(6), 74. https://doi.org/10.3390/ijgi5060074

Zook, M. (2017). Crowd-sourcing the smart city: Using big geosocial media metrics in urban governance. *Big Data & Society, 4*(1), 1–13. https://doi.org/10.1177/2053951717694384

Chapter 5
Transactional Participation: Evaluation from Local Government Perspective

Abstract The citizen-centric smart city is the latest and most comprehensive approach to use information technology to make local government operations more efficient and, ideally, more responsive to community needs. Citizen-centricity rests on the premise that residents are not passive consumers of digital data and services, but instead can choose how they use smart city services to participate more fully in society and to deal with the tangible urban problems they are confronted with. In this context, participation includes citizens' co-authoring of data and studies have shown how local governments can engage citizens to create spatiotemporal data about their surroundings, activities, and interactions with urban infrastructures. While dedicated apps can formalize how citizens collect data and communicate with local government, research shows that they can both limit the number of citizens who participate and can also accentuate existing inequities in participation. To date, the impacts of citizens having agency over tool choice remain unexplored from the perspective of local governments. We examine one aspect of this issue through the example of citizens using social media to report routine complaints and maintenance needs. Interviews were conducted with municipal government staff from 12 Canadian cities to document current challenges and practices of social media-based reporting. Results show that social media reporting shares many of the challenges other studies have noted for government use of participatory technologies; however, the quality of co-authored data and governments' limited capacity for technological change are seen as particularly important obstacles. Recommendations for developing internal workflows and strategies are provided for municipal governments to use social media as a channel for non-emergency reporting. More broadly, the results point to a need for clear assessments of how citizen centricity may be realized in evolving, and often patchy, smart city landscapes.

S. Zhang, *Geo-participation*, https://doi.org/10.1007/978-981-95-8228-0_5

5.1 Introduction

The concept of citizen-centric smart cities is often promoted as a comprehensive approach for using information technology to make local government operations both more efficient and more responsive to community and individuals' needs (Komninos et al., 2018; Mora et al., 2019). Whether this reorientation of smart city efforts toward citizens' needs, rather than the priorities of governments and technology firms, is primarily rhetoric or instead signals a fundamental shift in technology governance in cities remains a complex and contested issue (Cardullo & Kitchin, 2019a; Shelton & Lodato, 2019). In its best light, a citizen-centric smart city can provide its residents with new layers of affordances. These affordances allow citizens to engage meaningfully in defining the tangible urban problems of most concern to them, to delimit how technology should be applied to these problems, and to participate more effectively in society by leveraging the city's physical and digital assets (Jiang et al., 2022; Roche, 2016).

Evidence to date suggests that realizing citizen centricity will require several tensions to be reconciled or at least acknowledged explicitly. For instance, technology-led solutions to well-defined urban problems (e.g., real-time optimization of traffic flows) will continue to be more attractive to resource-constrained cities than bottom-up initiatives that have fewer direct ties to core city operations and take longer to show measurable results. Second, visions of seamlessly interoperable smart city systems accessible to all are at odds with the reality that readiness for using new technologies and data sources, "smart" or not, remains patchy and uneven. This patchiness exists geographically across and within cities, thematically in terms of the urban processes technologies are most feasible for, and socially in favor of technology-literate individuals who can best capitalize on new affordances (Jiang et al., 2022; Levenda et al., 2020). Finally, while citizen participation is often portrayed as critical to smart city success, technology both enhances and constrains these social processes (Berntzen & Johannessen, 2016). For instance, software that scales data access and communication (e.g., online meetings, surveys, mobile mapping, etc.) citywide also tends to channel engagement through few, standardized platforms which may not be accessible to all and often reduce participation to convenient, but often shallow, "clicks" on online surveys or maps (Cardullo & Kitchen, 2019b; Sweeting et al., 2022).

This study considers these tensions in light of citizen participation in smart cities being recast to include a range of passive (e.g., cellphone movement tracking) and active (e.g., citizen science) data authoring roles (Rodriguez Müller, 2021; Wu, 2020; Zhang, 2019). In this context, citizens' data allow local governments to better understand public needs, enable public service provision to be finetuned, and provide a source of free labor for resource-poor cities (Burns & Welker, 2022; Molinillo et al., 2019). For data authoring as participation to align with the ideals of citizen centrality, at a minimum individuals must have agency to choose what smart city processes they engage in (e.g., opting out of data collection) and how they engage with them (Sweeting et al., 2022). We explore cities' readiness to support one aspect of agency, namely individuals' choice of how they participate, through the example of citizens using social media to report non-emergency concerns and maintenance needs (e.g., graffiti, parking enforcement, noise complaints, potholes, etc.).

Many North American municipalities provide 311 services that allow citizens to easily access information about city operations (e.g., find residential waste pickup dates) on a 24/7 basis and to report incidents of concern through phone, email, website, and mobile applications (Nam & Pardo, 2014a; Sieber & Johnson, 2015). With respect to reporting, mobile applications offer advantages of independent and in situ use; however, they have low adoption rates due to the need for a smartphone, data plan, and a city-specific 311 app, some of which do not permit citizen-to-citizen sharing of reports (Desouza & Bhagwatwar, 2012; Lu, 2017; Peixoto & Fox, 2016). Previous studies show that social media has potential as a supplementary reporting mechanism since broader segments of the public already use social media, and these data often include important place-based information through embedded geographic coordinates, images, or text descriptions (Cho & Melisa, 2021; Feick & Robertson, 2015; Zhang & Feick, 2016;). Most local governments currently use social media widely for communication. However, as Cho and Melisa (2021) note, over 85% of this use is unidirectional and consists of local government informing and nudging citizens, while less than 9% of the interaction relates to bottom-up feedback and reporting.

In this chapter, we seek to identify key challenges that local governments in Canada encounter, and the resultant workflows they adopt, when citizens use social media to report routine concerns and maintenance needs. Two research questions are addressed based on a series of interviews with Canadian municipal government officials:

1. How do Canadian local municipalities currently handle social media reports? and
2. What challenges do local governments face in adopting social media for non-emergency reporting?

We believe that this topic has value on two levels. First, beyond its functions for citizen-government communication, social media can reveal citizen-sensed phenomena and events that may not be found in other spatial datasets, including those from official 311 apps (Zhang & Feick, 2016). Exploring why and how local governments use, or do not use, these data may uncover options for repurposing user-generated geographic information in city operations and also contribute to the broader discourse of the roles of participatory technologies. On a secondary level, most research to date points to modest progress toward citizen-centric smart cities with many initiatives being criticized as being more technology- than problem-focused (Cardullo & Kitchin, 2019a, b; Rodriguez Müller, 2021). Lessons from local governments' experiences leveraging social media for routine reporting may shed light on challenges that may arise when more complex technologies and social processes are integrated into what Shelton et al. (2015) aptly termed the "actually existing smart city."

5.2 Citizen Reporting and Social Media

Local governments provide important non-emergency services, handling issues like graffiti, parking enforcement, noise complaints and potholes, as the maintenance of city infrastructure and routine government services are their important responsibilities. Citizens requiring non-emergency city services are an important component of government–citizen interaction. Driven by developments in information technology and the large volume of non-emergency citizen requests (e.g., City of Toronto receives more than 3 million inquiries a year), local governments seek innovative government–citizen collaboration through which citizen input can easily be gathered (Wiseman, 2014). The 311 systems are an example of how information on public services and needs within neighborhoods can be gathered and redirected to citizens as they "make it easy for citizens to becoming involved in their community by simply picking up the phone to report a problem they see" (Phelan, 2010, p. 3). Growing numbers of North American municipalities have established 311 customer service centers and expanded the service to mobile applications that offer citizens easier access to the service 24/7 (Nam & Pardo, 2014a; Sieber & Johnson, 2015). These mobile applications foster e-participation and are reported to be helpful for specific groups of people that face difficulties using or having access to digital services.

Social media has the potential to make reporting more accessible to broader segments of the public, as it has a large user base and is widely adopted by local governments (e.g., 88% of local governments in the United States use social media) (Oliveira & Welch, 2013). However, it has not been much used for citizen reporting. According to the report of the City of Toronto, for the 3.6 M external 311 inquiries and 1.4 M "customer contacts," 85% were received by phone and only 3% by web and 8% by email. Studies have examined social media as a source of volunteered or ambient geographic information and suggested that it is increasingly attractive as a data source for urban analysis (Crooks et al., 2015). In particular, social media has great potential for understanding new elements of cities and thus could provide new insights for urban governance (Poorthuis et al., 2016). The expectation is that social media can serve as a potential means to allow for more citizen inputs and increase openness and accountability (Stamati et al., 2015). The most prominent use cases for this aspect of social media were in situations where real-time communications were paramount, such as emergency management. This ability to leverage social media during emergencies has been shown to be effective in other cases, including more general cases of crowdsourcing maps rather than social media (Zook et al., 2010). However, analyzing the world with geosocial media data may bring different challenges, such as the unstructured nature of much of the data making it difficult to conduct meaningful research. In practice, while some studies have examined the practice of government adopting social media and crowdsourcing for emergency management and discussed the associated barriers (Harrison & Johnson, 2019; Haworth, 2016), if and how social media is a useful tool for citizens reporting non-emergency issues remains underexplored.

5.3 Challenges in Local Government Adoption of Digital Participatory Technologies

Despite the wide use of social media as a communication channel, there are significant challenges associated with citizen co-producing data and service using social media. These challenges are not unique to social media, but are common in local government adoption of digital participatory technologies. That is, local governments "seem reluctant to engage and use such (digital participatory) platforms in their public policy and service delivery efforts" (Falco & Kleinhans, 2018). Here, challenges refer to any issues that may hinder local governments' use of technologies. Falco and Kleinhans (2018) categorize various challenges concerning local government adoption of digital participatory technologies into three main aspects, namely contextual, technological, and organizational (Fig. 5.1). Some of these challenges align with the three tensions faced by citizen-centric smart city as discussed earlier. In the following, we incorporate the three tensions with the conceptualization of Falco and Kleinhans (2018), and review challenges associated with local government of digital participatory technologies, with an emphasis on social media.

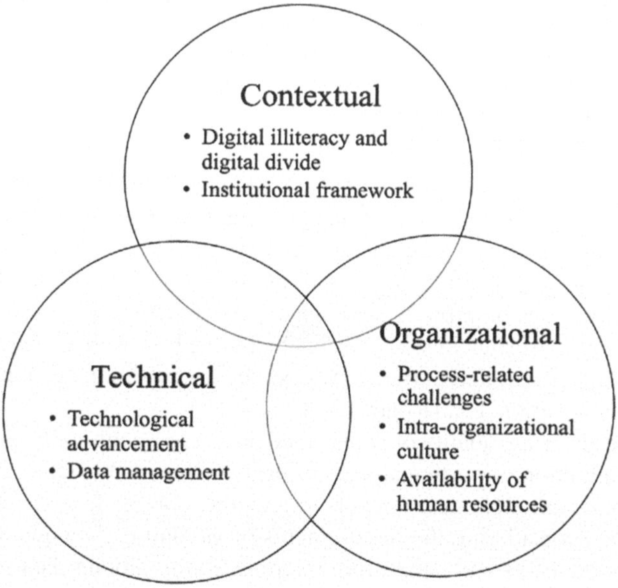

Fig. 5.1 Challenges of adopting digital participatory technologies by local governments

5.3.1 Contextual Challenges

Internet-based technologies introduce technical and social barriers to participation, as people need access to digital devices (e.g., computers and mobile devices) and the Internet, and must possess a certain level of digital and technical skills (Crutcher & Zook, 2009; Lin, 2013). Generational gaps are therefore observed in that younger generations are found to be more likely to accept digital participation methods (Corbett, 2013; Tulloch, 2008). The concern of excluding socially disadvantaged groups remains as these people may still lack necessary resources to digitally participate (Vicente & Novo, 2014).

A proper institutional framework to guarantee data privacy and security is another challenge faced by local government. Data collection through the Internet posits privacy concerns in that information about people's identity, communication, transaction, location and movement might be accessed, shared and used without people's knowledge or consent (Kitchin, 2016). The misuse of personally identifying information may raise unintended consequences such as safety concerns (Olteanu et al., 2016). From the citizen side, concerns of privacy may prevent people from contributing location data and result in scarcity or incompleteness of contributed data (Elwood & Leszczynski, 2013).

5.3.2 Technical Challenges

Technical challenges of local government adoption of digital participatory technologies are twofold. First, local governments face challenges of adapting to innovation and new technologies (Falco & Kleinhans, 2018). According to the well-established theory of diffusion of innovation, organizations' adoption of innovations is motivated by perceived advantage or improvement of using a new idea, approach or system over existing ones, yet is affected by other factors including how well an innovation aligns with existing procedures and norms, how easy the organization can understand, use and assess an innovation, etc. (Rogers et al., 2005).

Second, the uncertain quality of contributed data is one commonly identified challenge for the adoption of participatory platforms (Johnson & Sieber, 2013; Zhang, 2019). It is challenging for governments to integrate data with unassured quality to their operations, considering the requirements of validating data quality, managing heterogeneous data sets and supporting interoperability among data sets (Bakillah et al., 2013; Garnett & Kanaroglou, 2016). Moreover, the anonymity of online participation raises an issue of trustworthiness for governments to adopt contributed data (Johnson, 2016). Social media, in particular, is often characterized by unstructured content, anonymous authorship, lack of consistency, and biased representation (Elwood, 2010; Robertson & Feick, 2016; Sieber et al., 2016). The extent to which social media with unstructured content and inconsistent formats can fit into official non-emergency reporting programs is therefore worthy of further investigation.

5.3.3 Organizational Challenges

Making a digital innovation become a standard government practice should also account for a variety of administrative, policy, and institutional factors. Organizational constraints such as public agencies being more likely to trust their own capabilities of influencing participation, internal resistance to adopting new technology owing to factors such as job insecurity and fear of learning a new technology may obscure the incorporation of the new methods (Brown, 2012; Johnson & Sieber, 2013; Mergel, 2012). The inconformity of the new practice with existing procedures in an organization, the lack of relevant policy, and the cost associated with the use of the new technologies are also considered barriers to adopting new technologies in governments (Brabham, 2013; Haklay et al., 2014; Johnson & Sieber, 2013).

Despite the extensive studies on local government adoption of digital participatory technologies, few have explored what factors hinder the use of social media for non-emergency reporting. Do the above challenges pertain in the context of social media reporting? Among various aspects of challenges, what are the key obstacles faced by local governments? This study will use the conceptualization of Falco and Kleinhans (2018) as a foundation to further explore the challenges facing Canadian municipalities in adopting social media reporting. This exploration will help local governments evaluate the potential of using social media for improved service delivery and possibly make progress with the new practice based on experience and lessons learned from Canadian municipal governments.

5.4 Methods

5.4.1 Participant Recruitment

To gauge local governments' perspectives on citizens using social media for non-emergency reporting, interviews were requested with staff from 42 municipalities across Canada. To make the sample more representative of differences in municipal capacities, operational practices, and legislative contexts, sampling was stratified across provinces and by population size using the three classes (30,000–50,000, 50,000–150,000, 150,000 and over) suggested by El-diraby and Kinawy (2015). Municipalities smaller than 30,000 were excluded since smaller and/or rural municipalities often have resource constraints and responsibilities that differ from their larger urban counterparts (Seo & Bernsen, 2016) and will be considered in a forthcoming study. Staff involved with communication, 311 or customer-service, or operations departments were contacted using email addresses, social media accounts, or phone numbers listed on city web sites. In some instances, the person contacted recommended other staff with direct experience managing social media.

A total of 14 participants from 12 municipalities across 5 provinces were interviewed. From east to west, this group included participants from Newfoundland and

Labrador (St John's), Ontario (Burlington, Kitchener, Mississauga, Toronto, Stratford, Waterloo), Saskatchewan (Moose Jaw, Regina), Alberta (Calgary), and British Columbia (Langley, Vancouver). Nine interviewees were employed in a corporate services department, two were in citizen services, one in public works, and three in an information or digital service-related division. For the rest of the cities contacted, five indicated that they do not currently use social media reporting and thus would not participate. In-person interviews were conducted for four municipalities located near the researchers in southern Ontario, while the remainder were interviewed by phone.

5.4.2 *Interview Questions*

The interviews were designed around a mix of multiple-choice and open-ended questions organized into three main sections. First, interviewees were asked about their municipality's use of social media in general, including what social media platforms were being used and if and how social media was used for non-emergency reporting. The second section expanded on citizen reporting, with emphasis on current practices for processing and validating citizen reports and if their municipality saw more value in some citizen reporting methods than in others. The third section focused specifically on the challenges the interviewees encountered or anticipated with social media reporting. To aid responses, challenges commonly identified in the literature were provided. Interviewees were also encouraged to suggest additional challenges that were not listed.

5.4.3 *Thematic Analysis*

The interview results were examined using thematic analysis to identify and organize themes in the data (Braun & Clarke, 2006). As Nowell et al., (2017, p. 2) note, thematic analysis is particularly useful for "examining the perspectives of different research participants, highlighting similarities and differences, and generating unanticipated insights." Initial codes were identified in the transcripts by highlighting important sentences or paragraphs of text and assigning initial labels based on the authors' understanding of the text. Next, themes were identified from the notes using a deductive approach with challenges from previous studies serving as an initial framework for the main themes. The codes generated by authors in the previous step were assigned to the main themes. This step involved combining codes that belonged to the same theme, refining codes whose meaning was not clear initially, and separating codes which encompassed multiple meanings given the framework. No new themes emerged beyond those in the original framework through this process, although this can occur during this stage. In addition to assigning themes, connections between themes were developed as we found that related codes could belong to different themes. Finally, the theme and code collection were reviewed to ensure that each theme was internally coherent, was meaningful, and was supported by sufficient data (Braun & Clarke, 2006; Nowell et al., 2017). Some revisions to code and theme assignments were made as a result by repeating the previous steps.

5.5 Results

5.5.1 Current Practices of Social Media Reporting

Twitter (since X was named Twitter at the time of the interviews, we have elected to use the former name), now known as X, was used by all 12 surveyed municipalities with some having multiple accounts for different departments. Other popular social media used by municipalities were Instagram (7 municipalities), Flickr (5 municipalities), and YouTube (2 municipalities). Notably, all 12 municipalities had at least one staff member dedicated to publishing and updating social media content and communicating with citizens through social media.

With regard to citizen reports received from social media (primarily Twitter), participants' responses indicated that social media-based reporting was not accepted at four municipalities. In the remaining eight municipalities, social media-based reports were processed either by dedicated social media staff or by staff with broader customer service responsibilities. Figures 5.2–5.4 illustrate three main models that were identified from the interviews, referred to here as the redirect, stand-alone, and integrated models. Each model represents different workflows and organizational structures that the municipalities developed for using social media-based citizen reports, as well as other traditional sources such as email and phone calls. Table 5.1 provides an overview of the municipalities using each model and the approximate sizes of the populations they serve and their staff complements.

In the redirect model (Fig 5.2), municipalities do not use non-emergency reports received through social media. Instead, when non-emergency reports are received from social media accounts, customer service staff request via social media that citizen reporters resubmit their report through municipal phone call or email channels.

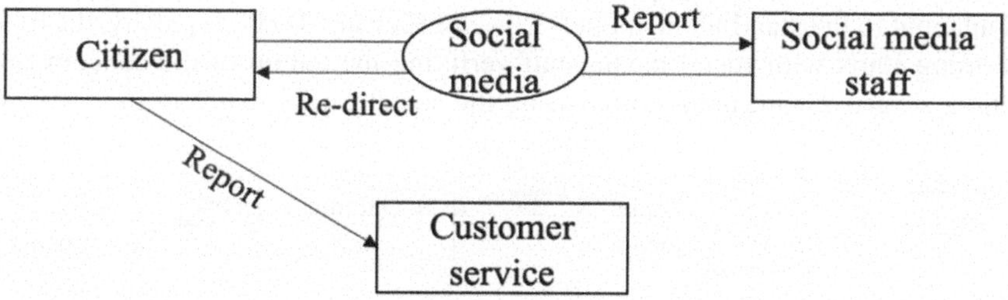

Fig. 5.2 The redirect model

Table 5.1 Characteristics of municipalities using each model

Redirect	A,M,O,L	30,000–100,000(3) >100,000 (1)	<400	NA
Stand-alone	C,(D,N),E,(I,J)	<100,000(1) 100,000–700,000(4)	800–1600(3); 160(1)	Corporate service (Communication division)
Integrated	B,G,H,K	<200,000(1) 600,000–2,600,000(3)	>9000(3); 800(1)	311 Citizen service, Enterprise services, Digital services, Records and information

Depending on the municipality's practices, customer service staff may act as an intermediary for citizen and municipal staff communication, or they may provide the citizen with contact details for the department(s) pertinent to the reported issue.

Four of the twelve interviewed municipalities used this redirect model and in all four cases social media acted as an intermediary. These municipalities had smaller populations, ranging from 30,000 to 190,000 (three cities under 100,000; one city has a population of 190,000), and less than 400 staff in total. In these municipalities, social media was considered mostly as a tool for outgoing communication flows to inform the public about government events, required notices (e.g., proposed planning bylaws), and other information. Government-citizen communications in these municipalities are handled predominantly by communication departments.

With the stand-alone model, social media-based reports are handled by dedicated social media staff housed in the corporate services unit. Unlike other citizen-government interactions (e.g., phone calls, emails) that predominantly flowed through a central customer service department, social media was seen as so distinct that it required staff with specialized skill sets. Staff from four interviewed municipalities used this model, three of which were mid- to large in size with populations between 100,000 and 700,000 and between 800 and 1600 total staff. One municipality using this approach, however, was considerably smaller in both population (<100,000) and staff complement (160).

In the stand-alone model (Fig. 5.3), social media is still seen as a special mode for citizen–government interactions; however, here it is used for two-way communication flows that are more participatory in nature than the purely communicative function inherent to the stand-alone model (Rowe & Frewer, 2005). Typically, the report processing starts with social media staff verifying the nature of the concern (e.g., location, severity) with the reporter using the social media platform the issue was

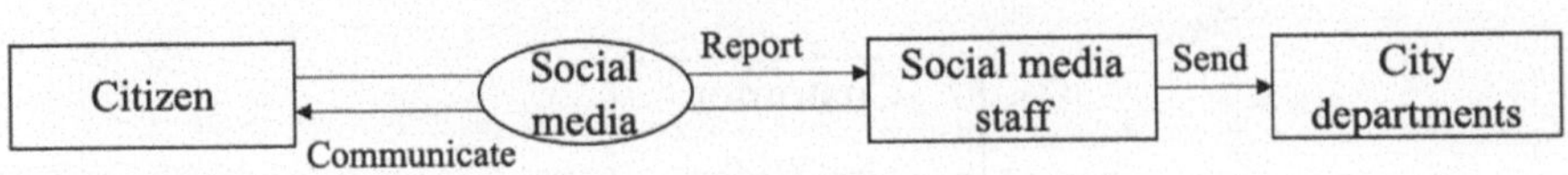

Fig. 5.3 The stand-alone model

Fig. 5.4 The integrated model

reported on. Next, social media staff liaise with the relevant department(s) to process the report. Any further inquiries and communication regarding the issue's resolution take place on the same social media platform. Among the four municipalities, two have used social media management software, such as HootSuite and Sprout, to help manage these two-way social media communication flows. The software can alert government staff when a social media message is directed to the municipality and manage communication history to help staff track earlier conversations (participants E,I,J).

In contrast to the Stand-alone model, some municipalities channel all modes of citizen reporting (e.g., phone calls, social media, emails) through centralized customer service or non-emergency 311 units. This Integrated model (Fig. 5.4) was used by three of the larger municipalities (populations between 600,000 and 2,600,000, more than 9000 staff) as well as one mid-sized municipality (population of 190,000, 800 staff). These municipalities have well-developed customer relationship management (CRM) systems (e.g., Microsoft Dynamics) which allow staff to collect and relate citizen reports from multiple channels and to easily communicate with citizens concerning any missing information and/or progress on resolving the issue. Internally, staff send the report to relevant departments if necessary and assign a central case number to the report for further inquiry once the report is validated. While the departments handling social media reports are differently named in these four municipalities, they all have information or digital service-related divisions for technical support.

5.5.2 *Challenges of Adopting Social Media Reporting*

We divided interviewed municipalities into three groups based on how they currently handle social media reporting. Results show that major challenges facing participating Canadian municipalities are technological challenges and organizational constraints. However, the three groups perceive the challenges differently (Table 5.2) and accordingly adopt different strategies to tackle with the challenges.

Contextual Challenges

Table 5.2 Challenges of adopting social media reporting faced by different municipalities

Challenges		Redirect	Stand-alone	Integrated
Contextual	Digital literacy	–	–	–
	Privacy	–	–	–
Technological	Data quality	□	□	□
	Trustworthiness and credibility	□	+	□
	Technological advancement	×	–	–
Organizational	Process-related challenges	×	□	□
	Intra-organizational culture	×	+	+
	Cost	×	□	+

Digital Literacy

All of the interviewed municipalities consider that digital literacy is not an issue for adopting social media reporting. Rather, social media have already been widely used by the public. For municipal governments, they accept social media as a reporting method only because it is preferred by the public.

Privacy

Privacy is not a major concern for municipalities to use social media posts, as interviewees believe that the information is already publicly available once social media posts are published online. That is, an "implicit consent" (Harrison & Johnson, 2019) is agreed when a citizen voluntarily contributes information. Government staff would communicate with reporters through private messages when they need to collect further information, especially identifiable information, from citizens, so that no information would be disclosed without the consent of the citizen.

However, interviewees suggest that while privacy is not an issue for government adopting social media, it may prevent local citizens from using social media reporting. Eight out of the fourteen interviewees believe that citizens may have concern that posting on social media would intrude their personal privacy and will unlikely disclose their personal information such as home address. Other four interviewees have contradictory views in that they believe that citizens, especially those who are already social media users, would understand the context of using social media and would not have issues with disclosing certain information such as a public location where an issue happens. Similar to what is suggested by Ricker et al. (2015), participant L believes that privacy will be less of a concern when citizens perceive benefits outweigh potential risks. Rather, it is the lack of knowledge about privacy that may prevent people from using online participatory services (Christin et al., 2011).

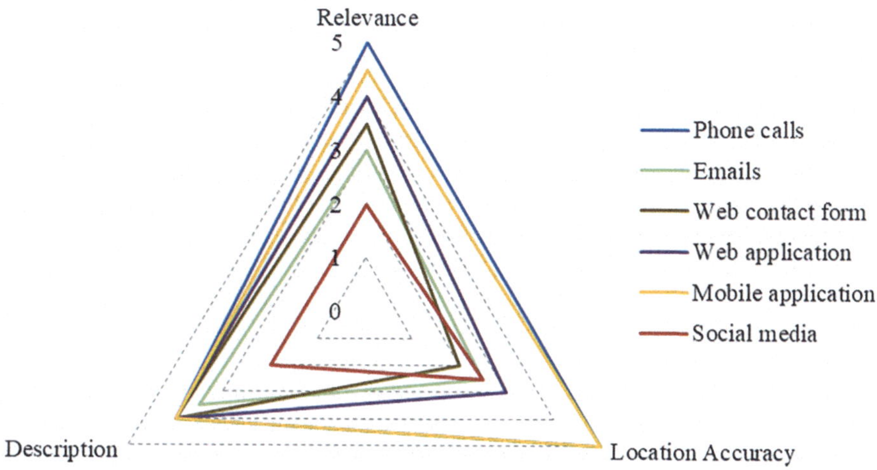

Fig. 5.5 Comparison of the quality of citizen reports collected from different channels

The privacy concerns are also considered to vary among different generations. Participant G suggests that the younger generation who are used to share their lives on social media would have less privacy concerns compared to their elder counterparts.

Technological Challenges

Data Quality

Municipalities, whether adopt social media reporting or not, perceive data quality as a major challenge. The interviewees were asked to evaluate the quality of citizen reports collected from six primary channels of non-emergency reporting. In particular, data quality aspects including relevance, description (semantic accuracy), and locational accuracy are evaluated, as they are the most-addressed quality aspects of contributed spatial data in the literature (Olteanu-Raimond et al., 2017). Figure 5.5 shows the average scores interviewees gave to each of the three aspects of citizen reports from each reporting channel. Overall, the quality of social media reporting was considered the poorest, in that all three quality aspects of social media reports are lower than those of non-emergency reports collected through other channels (Fig. 5.5).

Relevance

Interviewees suggested that many of the social media comments are only general complaints that do not reflect on actual problem that requires attention. As indicated by Participant B:

> They (social media messages) are not specific about what the issue is. And then sometimes people just want to 'chill' the city, they want to make comments or be provocative.

Moreover, interviewees raised the issue that social media conversations often go beyond the scope of the original issue. According to participants C and D, the open and continuing dialogs on social media make it difficult for government staff to track the origin of an issue and decide whether the issue is relevant to the jurisdiction of the municipality or individual department.

Description

Interviewed staff suggested that unlike web and mobile reporting applications where people answer a list of pre-defined questions when reporting an issue, people use their own words to describe an issue when using social media. Yet, the descriptions often do not provide enough details that are necessary for government operations, according to interviewees. Moreover, there is the issue of semantic ambiguity that different people may describe the same issue differently. Rather, all of the interviewees considered photos as better representatives of real-world circumstances.

Locational Accuracy

Geolocation is an important component of non-emergency reporting. In a separate question about how much social media reports have implicit or explicit geographic information, all the interviewees indicated that only less than 5% of social media reports have exact geographic information such as geotags and addresses. Therefore, municipalities are not collecting or using geotags associated with social media reports. Instead, they would identify the location of an issue based on reporter's description. Yet, the description is often ambiguous. Participant E gave an example of people referring to an intersection of two streets even though those streets cross several times along their length and have multiple intersections. Participant C also pointed out that different locations with same or similar names make it difficult for government staff to determine where the problem is and whether it is within the jurisdiction. According to participants I and J, the staff often need to use maps and their own knowledge to validate the location.

Preferably, the automatic geotagging function of social media can be used similarly to location services embedded in web and mobile applications. All of the interviewees agreed that geotagging and reverse geocoding (i.e., matching an input address to a map location) functions could help improve locational accuracy of reports. Yet, two questions were brought up. First, when people use their current location to report an issue, this location may not be the location of the issue. Participant L gave an example of reporting potholes where the reporter is unlikely to stand at the pothole location when geotagging. Participant I also pointed out that the GPS of a mobile device is often not sufficiently accurate for operators to find out the location of a reported issue. Second, current mapping or geotagging approaches only provide the staff with location information (x, y) but not height information (z). According to participant K, it is time-consuming for staff to locate a reported issue that happens at a multi-story building without the height information.

Trustworthiness and Credibility

Because of the quality issues, the redirect group and integrated group consider trustworthiness and credibility issues as a major challenge of social media reporting. As participant A points out:

> … there is a concern that if we send our staff out but we can't find it because it wasn't geotagged or, you know, we didn't get to ask them the questions of exactly what it was, then we waste staff resources if we assign staff based on social media.

The anonymity of social media users further exacerbated such concern. Participants A and D claimed that not having contact information of reporters makes further inquiry and validation of a reported issue more difficult. Moreover, participants A and E raised the concern that people may say whatever they want to say behind the veil of anonymity and send negative messages that do not always hold the truth. Participants D, H, and K had similar observations that some frequent reporters repeatedly complain about the same or similar issues. Staff have to rely on their personal knowledge or "common sense" to determine the validity of a report. For instance, the staff often need to look up a reporter's social media account to ensure the credibility of the reporter. This is similar to the findings with general online content. That is, determining the trustworthiness of online content is mostly subjective (Banerjee et al., 2017).

Some municipalities from stand-alone group have contradictory views. Participant E believes that tracking the histories of an individual's social media records makes social media more reliable. That is, government staff could comprehensively assess the legitimacy and credibility of the report by putting together information including the individual's biography information, pictures, the number of followers, and past interactions.

Technological Advancement

For redirect group, their limited technical capability is a major discouraging factor that prevents them from adopting social media reporting. In particular, interviewed municipal governments consider that they do not have the technology required to manage social media messages. For stand-alone and integrated groups, technological advancement is not a major concern. Most of these municipalities have experience with adapting to new technologies (e.g., CRM system). Therefore, they are more confident with or have transferable skills and procedures for adopting technological innovation. In our interviews, four interviewed municipalities have already deployed social media management software to improve the efficiency of social media communication. The technical capacities for tackling citizen reporting from various channels also assist with handling social media reports. According to participant H, the municipality has been working on automated and integrated methods of pre-processing and validating citizen reports, as they could improve the efficiency of municipalities' customer service and reduce the risk of generating errors by manually handling citizen reports. These methods can be generalized to social media reporting.

Organizational Challenges

Process-Related Challenges

Process-related challenges are considered as a discouraging factor for the redirect group and a major challenge for the stand-alone and the integrated groups. Some of the process-related challenges of handling social media reports relate to the quality issues of social media reports. Local government often need specific parcel information for handling service request that is often not provided by public map service (e.g., Google Maps). As a result, government staff need to take additional steps to match user-provided location information with parcel information used in government systems. According to participant B, departments need to drill down and ask a series of questions to ensure all the right information is received. The simplicity and often incompleteness of social media messages may make the staff nervous because of the uncertainties. Relevantly, the municipal governments the redirect group have little desire to formalize the process of social media reporting, considering the complexity of validation, but stick with the traditional reporting methods (Participants A,M,O).

With the stand-alone group, an additional challenge is the lack of formalized guidelines. According to participant D, their staff are sometimes not sure to which department they should forward the issue, and what information the other department would need to handle the issue. Lack of an integrated management system or database may also result in repeated work within governments as whether a reported issue has already been handled is not automatically tracked.

Intra-Organizational Culture

Intra-organizational culture is considered a discouraging factor by the redirect group. In addition to technological challenges that have increased the barriers of accepting new technologies within municipalities, participants consider social media to be a valuable citizen-engagement tool rather than a reporting channel. They do not consider that the investment for adopting social media would outweigh the value brought by social media reporting (participants A,M,O). Some municipalities in the other two groups agree with this viewpoint. While participants acknowledge that social media would improve communication between governments and citizens and enhance the transparency of municipal governments (Picazo-Vela et al., 2012), they do not think there is a huge demand from the public, especially in the cities where the proportion of aging population is significant in the total population (participant C).

Mixed views are revealed in stand-alone and integrated groups. For some municipalities, the adoption of social media is mainly driven by its wide use by the public and an absence of independent 311 tools (participant E). As more citizens use social media for reporting, the municipalities begin to allocate resources to handle social media reporting and gradually formalize procedures for processing social media reports. In other words, the accepting culture within an organization could inform policy changes and further promote the adoption of social media reporting (Harrison & Johnson, 2019). In other cases, municipalities' experiences with social media communication and mobile reporting applications could increase the acceptance level of social media reporting in an organization. Interviewees state that they would like to promote social media reporting based on their previous experiences with mobile reporting applications and other pilot projects of automated customer services (participants G,I,L). The previous experiences could also make the benefits of a technological innovation more visible to municipal governments, so that municipalities are more willing to adopt the new reporting method or at least test the new technology with pilot projects.

Cost

Related to the poor quality of social media reports, extra work of validating citizen reports is required for further processing through work orders and permit requests. For web and mobile applications, they usually require citizens to fill out a reporting form that asks for relevant information such as the location and the type of a reported issue. This guided reporting procedure acts as a form of quality control and results in good-quality reports that can be directly integrated into back-end systems (e.g., CRM system) for further processing (Foth et al., 2011; Lu, 2017). Comparatively, there is no such procedure for social media. Validating social media reports is primarily based on government staff's communication with citizen reporters and staff's personal knowledge. For example, the staff would ask the type and the location of the issue if certain information is not clear in the report. Moreover, the communication through private messages of social media is not always efficient. Government staff often need to manually search the communication history to find the reporter for follow-ups, yet the reporter may not always respond timely.

Considering the high cost of validating social media reports, municipalities in the redirect group state that they do not have the capacity (i.e., financial and human resources) to process citizen requests from social media. For stand-alone and integrated groups, municipal governments are also struggling with extra time and human capital needed to validate and process social media reports. According to participant B, their staff may spend twice as much as the time and effort needed for reports from other channels. An additional challenge arises regarding the tension between citizen expectation of real-time response, resulting from the convenience of social media for reporting, and high costs of handling citizen requests.

5.6 Discussion

5.6.1 Key Obstacles

The study shows that Canadian municipalities face critical technological and organizational challenges for adopting social media reporting. The identified technological and organizational challenges are not unique to social media reporting, but are relatively common in local governments' adoption of new technologies. However, interview results suggest that identified challenges are related and that data quality and technological advancement are key obstacles faced by municipalities (Fig. 5.6). That is, these two factors are not the outcomes of others, but can cause other obstacles. The ability of local governments to adapt to new technologies affects the extent to which a local government accepts social media reporting as well as the formation of procedures for handling social media reports. Trustworthiness and credibility are outcomes of poor data quality of social media reports, as are the cost and procedures for handling social media reports. Issues of credibility and cost further impact intra-organizational culture regarding municipalities' attitudes toward social media reporting. Additionally, issues of trustworthiness and credibility exacerbate process-related challenges, as additional procedures are required by municipal governments to validate social media reports. This results in an increase in human and financial costs of handling social media reporting.

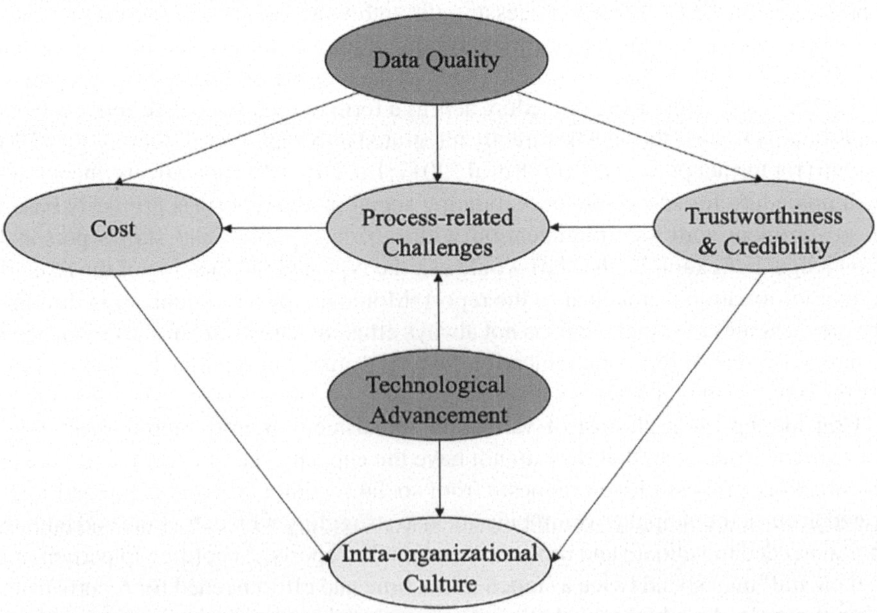

Fig. 5.6 The relationships among identified challenges of adopting social media reporting

Our results suggest that technological advancement is the key discouraging factor that prevents municipalities in the redirect group from using social media reporting. All the other discouraging factors are, to some extent, consequences of technological advancement. In our interviews, some small-sized municipal governments have quite limited technical capabilities. They are intimidated by the poor quality and incredibility of social media, as they would not have the human resources nor technical capability to handle the reports. As a result, they chose not to include social media as a citizen reporting channel.

Comparatively, technological advancement is not a major concern for municipalities that have adopted social media reporting, as these municipalities have legacy IT and customer service systems with which they can process social media reports. These municipalities face challenges arising from the quality of social media reports. Lack of an effective procedure for processing social media reports as well as the high costs of handling social media reports, in particular, is a common obstacle encountered by municipal governments. Despite the challenges, municipal governments have developed different strategies considering their own legacy systems and technical capabilities. Some mixed views are revealed: a challenge encountered by one municipality may have been, to some extent, solved by another. As such, suggestions can be drawn to shed light on strategies for adopting social media reporting.

5.6.2 Suggestions for Adopting Social Media Reporting

Despite the challenges, 12 out of 14 interviewees acknowledged the importance of social media as a communication channel and believed that municipalities should adopt reporting methods favored by the public. People are more likely to report with social media immediately after they see the issue, whereas the likelihood of people seeing the issue and reporting later using other methods is much lower (participants D,E,G,L). Some participants suggest that local governments should take the main responsibility of handling unstructured citizen reports and encourage citizens to report in their preferred methods, instead of asking citizens to go through multiple steps to report or restricting citizens to specific reporting channels (participants E,G,H,L). To better take advantage of social media reporting, it is critical to focus particularly on challenges related to the quality of social media reports and technological advancement of municipal governments.

First, the mismatch between organizational needs on precise information and unassured quality of citizen contribution is not unique to social media but is found common with volunteered citizen reporting (Brandeis & Nyerges, 2016). That is, governments often need accurate and often specific information (e.g., parcel information and height information) to ensure the efficiency and lower the cost of government operations. Training the public is often recommended to address this gap. For example, empirical studies suggest that the data quality of non-expert contributions for land use mapping may evolve over time and eventually be close to that of expert contributions (See et al., 2013). It is recommended to set up protocols and systems in advance to promote VGI practices (Haworth, 2016). In the context of social media reporting, it could be beneficial for municipalities to establish formal procedures and standards for social media reporting and guide users with recommended reporting routines (e.g., reporting format and content). As stated by some of the interviewees, it is important to provide users with guidance on how and what they should report (participants A,G). For instance, geotagging services, albeit the hardware deficiency, are increasingly used and considered beneficial for locating a reported issue. However, the public's low adoption rate of the tool could be related to their lack of knowledge of the tool and the benefits and risks associated with using it (participant L). It is therefore beneficial to educate the public and increase public awareness of the geospatial tools and methods (participants H,L).

Second, to strengthen the benefits of social media for communication and participation, municipal governments should leverage organizational capacities and establish clear strategies of adopting social media for particular uses such as citizen reporting (Oliveira & Welch, 2013). Our study suggests that developing an integrated program of municipal reporting could improve government efficiency, as municipalities that deploy the integrated model of citizen reporting have fewer issues of unclear or overlapped responsibilities of multiple departments. For municipalities that already have an integrated 311 reporting program or equivalent customer service programs, it could be beneficial to develop procedures for processing social media reports and make it compatible with the existing program. However, creating service integration capabilities could be challenging considering the resources and multi-department efforts for coordination (Nam & Pardo, 2014b). For municipalities that do not have enough human and technical resources, it might be better for them to stick with the redirect model of utilizing social media reporting and to start building a partnership with neighboring jurisdictions and third-party agencies (Harrison & Johnson, 2019).

Additionally, it is unlikely that all local governments would take revolutionary shifts in how they operate. Rather, some municipalities may adapt to new technologies sooner because of public pressure or internal needs of optimizing government resources, whereas others may take evolutionary steps of exploring and adopting new technologies. In our interviews, some of the municipalities that have not used geosocial media reporting expressed the desire of learning from others' experiences to determine whether they would adopt geosocial media reporting or not. According to Rogers (1983), networking among adopters may help reduce the uncertainty of innovation adoption and help late adopters to feel safer about adopting innovation. It is therefore helpful for cities that led technological changes to share their experiences of implementing technology-driven initiatives so that other cities could better understand the processes and outcomes of implementing an innovation (Johnson & Sieber, 2013).

5.7 Conclusion

While social media have become a critical tool for local governments to communicate with citizens, few scientific studies have explored their use for non-emergency reporting. Through a series of interviews with Canadian municipal governments, this study brings a Canadian perspective to the current practices of adopting social media reporting and associated challenges. Three models of adopting social media reporting are identified in Canadian municipalities, including redirect, stand-alone, and integrated models. The adoption of models is associated with the sizes and technical capacities of municipal governments. Particularly, we found that lack of technical capability is the key obstacle that prevents small-sized municipalities from adopting social media reporting. Comparatively, data quality and, accordingly, process-related challenges and human and financial costs are the key challenges faced by municipalities which have accepted social media as a reporting channel. Contextual challenges, including digital literacy and privacy issues, are not major concerns for municipalities to adopt social media reporting.

In addition to the understanding of the current practices of adopting social media for non-emergency reporting in Canadian municipalities, this study provides valuable insights into developing social media-based initiatives in local governments. The practice of integrating social media with 311 or customer service programs in some interviewed municipalities demonstrates the possibility of overcoming some of the challenges and sheds light on managing citizen reports from various channels. With increasing use of citizen reporting apps, governments are overwhelmed with increasing demands from citizens (Desouza & Bhagwatwar, 2012). In the broad discourse of achieving citizen-centric smart city approaches, the approach of integrating customer service programs, together with developments in automated report processing systems, is not specific to social media reporting, but will be useful for escalating government capacities of handling increasing citizen demands.

Despite the contributions, this study has several limitations that should be addressed in future research. First, this chapter suggests service integration for handling citizen reports from multiple channels. Although some recommendations regarding technological and organizational developments for service integration are provided based on the interviews, practicing service integration involves other organizational and cross-organizational factors (Nam & Pardo, 2014b) that need to be further understood. Second, the study identifies quality and credibility challenges associated with social media reporting based on government perspectives, but how citizens value social media reporting and evaluate the quality of social media information is yet to be known. It is worthwhile to further compare local government perspective and citizen perspective to assess the value of social media for non-emergency reporting and to develop strategies for the adoption of social media. Third, the study is limited to a sample of English-speaking municipalities in Canada. The division of French-speaking and English-speaking regions in Canada may raise issues related to availability of information in different languages (Bertot et al., 2012), which is worth further investigation.

References

Bakillah, M., Liang, S., Zipf, A., & Arsanjani, J. (2013). Semantic interoperability of sensor data with volunteered geographic information: A unified model. *ISPRS International Journal of Geo-Information, 2,* 766–796. https://doi.org/10.3390/ijgi2030766

Banerjee, S., Bhattacharyya, S., & Bose, I. (2017). Whose online reviews to trust? Understanding reviewer trustworthiness and its impact on business. *Decision Support Systems, 96,* 17–26. https://doi.org/10.1016/j.dss.2017.01.006

Berntzen, L., & Johannessen, M. R. (2016). The role of citizen participation in municipal smart city projects: Lessons learned from Norway. In *Smarter as the New Urban Agenda* (pp. 299–314). https://doi.org/10.1007/978-3-319-17620-8

Bertot, J. C., Jaeger, P. T., & Hansen, D. (2012). The impact of polices on government social media usage: Issues, challenges, and recommendations. *Government Information Quarterly, 29*(1), 30–40. https://doi.org/10.1016/j.giq.2011.04.004

Brabham, D. C. (2013). *Using crowdsourcing in government.* IBM Center for The Business of Government.

Brandeis, M. W., & Nyerges, T. (2016). Assessing resistance to volunteered geographic information reporting within local government. *Transactions in GIS, 20*(2), 203–220. https://doi.org/10.1111/tgis.12168

Braun, V., & Clarke, V. (2006). Qualitative Research in Psychology Using thematic analysis in psychology Using thematic analysis in psychology. *Qualitative Research in Psychology, 3*(2), 77–101. http://www.tandfonline.com/action/journalInformation?journalCode=uqrp20%5Cn

Brown, G. (2012). Public participation GIS (PPGIS) for regional and Environmental Planning: reflections on a decade of empirical research. *URISA Journal, 25*(2), 7–18.

Burns, R., & Welker, P. (2022). "Make our communities better through data": The moral economy of smart city labor. *Big Data & Society, 9*(1), 20539517221106380. https://doi.org/10.1177/20539517221106381

Cardullo, P., & Kitchin, R. (2019a). Being a 'citizen' in the smart city: Up and down the scaffold of smart citizen participation in Dublin, Ireland. *GeoJournal, 84*(1), 1–13. https://doi.org/10.1007/s10708-018-9845-8

Cardullo, P., & Kitchin, R. (2019b). Smart urbanism and smart citizenship: The neoliberal logic of 'citizen-focused' smart cities in Europe. *Environment and Planning c: Politics and Space, 37*(5), 813–830. https://doi.org/10.1177/0263774X18806508

Cho, W., & Melisa, W. D. (2021). Citizen coproduction and social media communication: Delivering a municipal government's urban services through digital participation. *Administrative Sciences, 11*(2). https://doi.org/10.3390/ADMSCI11020059

Christin, D., Reinhardt, A., Kanhere, S. S., & Hollick, M. (2011). A survey on privacy in mobile participatory sensing applications. *Journal of Systems and Software, 84*(11), 1928–1946. https://doi.org/10.1016/j.jss.2011.06.073

Corbett, J. (2013). "I don't come from anywhere": Exploring the role of the Geoweb and volunteered geographic information in rediscovering a sense of place in a dispersed Aboriginal community. In D. Sui, S. Elwood, & M. Goodchild (Eds.), *Crowdsourcing geographic knowledge: Volunteered geographic information (VGI) in theory and practice* (pp. 223–241). Springer. https://doi.org/10.1007/978-94-007-4587-2

Crooks, A., Pfoser, D., Jenkins, A., Croitoru, A., Stefanidis, A., Smith, D., ... Lamprianidis, G. (2015). Crowdsourcing urban form and function. *International Journal of Geographical Information Science, 29*(5), 720–741. https://doi.org/10.1080/13658816.2014.977905

Crutcher, M., & Zook, M. (2009). Placemarks and waterlines: Racialized cyberscapes in post-Katrina Google Earth. *Geoforum, 40*(4), 523–534. https://doi.org/10.1016/j.geoforum.2009.01.003

Desouza, K. C., & Bhagwatwar, A. (2012). Citizen apps to solve complex urban problems. *Journal of Urban Technology, 19*(3), 107–136.

El-diraby, T. E., & Kinawy, S. N. (2015, March). *Analyzing approaches used by Ontario municipalities to develop road asset management plans: Initial insights report*. Residential and Civil Construction Alliance of Ontario.

Elwood, S. (2010). Geographic information science: Emerging research on the societal implications of the geospatial web. *Progress in Human Geography, 34*(3), 349–357. https://doi.org/10.1177/0309132509340711

Elwood, S., & Leszczynski, A. (2013). New spatial media, new knowledge politics. *Transactions of the Institute of British Geographers, 38*(4), 544–559. https://doi.org/10.1111/j.1475-5661.2012.00543.x

Falco, E., & Kleinhans, R. (2018). Beyond technology: Identifying local government challenges for using digital platforms for citizen engagement. *International Journal of Information Management, 40*(December 2017), 17–20. https://doi.org/10.1016/j.ijinfomgt.2018.01.007

Feick, R., & Robertson, C. (2015). A multi-scale approach to exploring urban places in geotagged photographs. *Computers, environment and urban systems, 53*, 96–109.

Foth, M., Schroeter, R., & Anastasiu, I. (2011). Fixing the city one photo at a time: Mobile logging of maintenance requests. In *Proceedings of the 22nd Australasian Conference on Computer-Human Interaction* (pp. 126–129). ACM. https://doi.org/10.1145/2071536.2071555

Garnett, R., & Kanaroglou, P. (2016). Qualitative GIS: An open framework using SpatiaLite and open source GIS. *Transactions in GIS, 20*(1), 144–159. https://doi.org/10.1111/tgis.12163

Haklay, M., Antoniou, V., & Basiouka, S. (2014). *Crowdsourced Geographic Information Use in Government*. 1–79. http://discovery.ucl.ac.uk/1433169/

Harrison, S., & Johnson, P. (2019). Challenges in the adoption of crisis crowdsourcing and social media in Canadian emergency management. *Government Information Quarterly, 36*(3), 501–509. https://doi.org/10.1016/j.giq.2019.04.002

Haworth, B. (2016). Emergency management perspectives on volunteered geographic information: Opportunities, challenges and change. *Computers, Environment and Urban Systems, 57*, 189–198. https://doi.org/10.1016/j.compenvurbsys.2016.02.009

Jiang, H., Geertman, S., & Witte, P. (2022). Smart urban governance: An alternative to technocratic "smartness." *GeoJournal, 87*(3), 1639–1655. https://doi.org/10.1007/s10708-020-10326-w

Johnson, P. A. (2016). Models of direct editing of government spatial data: Challenges and constraints to the acceptance of contributed data. *Cartography and Geographic Information Science, 0406*(May), 1–11. https://doi.org/10.1080/15230406.2016.1176536

Johnson, P. A., & Sieber, R. E. (2013). Situating the adoption of VGI by government. In D. Sui, S. Elwood, & M. Goodchild (Eds.), *Crowdsourcing Geographic Knowledge: Volunteered Geographic Information (VGI) in Theory and Practice* (pp. 65–81). Springer. https://doi.org/10.1007/978-94-007-4587-2

Kitchin, R. (2016). *Getting smarter about smart cities: Improving data privacy and data security*. Data Protection Commissioner of Ireland.

Komninos, N., Kakderi, C., Panori, A., & Tsarchopoulos, P. (2018). Smart city planning from an evolutionary perspective. *Journal of Urban Technology*, 1–18. https://doi.org/10.1080/10630732.2018.1485368

Levenda, A. M., Keough, N., Rock, M., & Miller, B. (2020). Rethinking public participation in the smart city. *The Canadian Geographer/le Géographe Canadien, 64*(3), 344–358. https://doi.org/10.1111/cag.12601

Lin, W. (2013). Situating performative neogeography: Tracing, mapping, and performing "Everyone's East Lake." *Environment and Planning A, 45*(1), 37–54. https://doi.org/10.1068/a45161

Lu, Q. (2017). *Smartphone applications in government: Characterizing and evaluating municipal smartphone applications for service requests* [University of Waterloo]. UWSpace. https://uwspace.uwaterloo.ca/bitstream/handle/10012/11151/Lu_Qing.pdf?sequence=1&isAllowed=y

Mergel, I. (2012). The social media innovation challenge in the public sector. *Information Polity, 17*(3–4), 281–292. https://doi.org/10.3233/IP-2012-000281

Molinillo, S., Anaya-Sánchez, R., Morrison, A. M., & Coca-Stefaniak, J. A. (2019). Smart city communication via social media: Analysing residents' and visitors' engagement. *Cities, 94*(December 2018), 247–255. https://doi.org/10.1016/j.cities.2019.06.003

Mora, L., Deakin, M., & Reid, A. (2019). Combining co-citation clustering and text-based analysis to reveal the main development paths of smart cities. *Technological Forecasting and Social Change, 142*(July 2018), 56–69. https://doi.org/10.1016/j.techfore.2018.07.019

Nam, T., & Pardo, T. A. (2014a). The changing face of a city government: A case study of Philly311. *Government Information Quarterly, 31*(SUPPL.1), S1–S9. https://doi.org/10.1016/j.giq.2014.01.002

Nam, T., & Pardo, T. A. (2014b). Understanding Municipal Service Integration: An Exploratory Study of 311 Contact Centers. *Journal of Urban Technology, 21*(1), 57–78. https://doi.org/10.1080/10630732.2014.887933

Nowell, L. S., Norris, J. M., White, D. E., & Moules, N. J. (2017). Thematic analysis: Striving to meet the trustworthiness criteria. *International Journal of Qualitative Methods, 16*(1), 1–13. https://doi.org/10.1177/1609406917733847

Oliveira, G. H. M., & Welch, E. W. (2013). Social media use in local government: Linkage of technology, task, and organizational context. *Government Information Quarterly, 30*(4), 397–405. https://doi.org/10.1016/j.giq.2013.05.019

Olteanu, A., Castillo, C., Diaz, F., & Kiciman, E. (2016). Social data: Biases, methodological pitfalls, and ethical boundaries. *SSRN Electronic Journal, 1–47*,. https://doi.org/10.2139/ssrn.2886526

Olteanu-Raimond, A.-M., Hart, G., Foody, G. M., Touya, G., Kellenberger, T., & Demetriou, D. (2017). The scale of VGI in map production: A perspective on European national mapping agencies. *Transactions in GIS, 21*, 74–90. https://doi.org/10.1111/tgis.12189

Peixoto, T., & Fox, J. (2016). When does ICT-enabled citizen voice lead to government responsiveness? *IDS Bulletin, 41*(1), 23–39. https://doi.org/10.19088/1968-2016.104

Phelan, A. (2010). Access Cupertino: Citizen engagement for the 21st century. ICMA.

Picazo-Vela, S., Gutiérrez-Martínez, I., & Luna-Reyes, L. F. (2012). Understanding risks, benefits, and strategic alternatives of social media applications in the public sector. *Government Information Quarterly, 29*(4), 504–511. https://doi.org/10.1016/j.giq.2012.07.002

Poorthuis, A., Zook, M., Taylor Shelton, M. G., & Stephens, M. (2016). 16 Using Geotagged Digital Social Data in. *Key methods in geography*, 248.

Ricker, B., Schuurman, N., & Kessler, F. (2015). Implications of smartphone usage on privacy and spatial cognition: Academic literature and public perceptions. *GeoJournal, 80*(5), 637–652. https://doi.org/10.1007/s10708-014-9568-4

Robertson, C., & Feick, R. (2016). Bumps and bruises in the digital skins of cities: Unevenly distributed user-generated content across US urban areas. *Cartography and Geographic Information Science, 43*(4), 283–300. https://doi.org/10.1080/15230406.2015.1088801

Roche, S. (2016). Geographic information science III: Spatial thinking, interfaces and algorithmic urban places-toward smart cities. *Progress in Human Geography, 0309132516650352*,. https://doi.org/10.1177/0309132516650352

Rodriguez Müller, A. P. (2021). Making smart cities "Smarter" through ICT-enabled citizen coproduction. In *Handbook of Smart Cities* (pp. 539–559). https://doi.org/10.1007/978-3-030-69698-6_63

Rogers, E. M. (1983). *Diffusion of innovations* (3rd ed.). Free Press.

Rogers, E. M., Medina, U. E., Rivera, M. A., & Wiley, C. J. (2005). Complex adaptative systems and the diffusion of innovations. *The Innovation Journal, 10*(3), 1–26.

Rowe, G., & Frewer, L. J. (2005). A typology of public engagement mechanisms. *Science, Technology & Human Values, 30*(2), 251–290. https://doi.org/10.1177/0162243904271724

See, L., Comber, A., Salk, C., Fritz, S., van der Velde, M., Perger, C., Schill, C., McCallum, I., Kraxner, F., & Obersteiner, M. (2013). Comparing the quality of crowdsourced data contributed by expert and non-experts. *PLoS ONE, 8*(7), 1–11. https://doi.org/10.1371/journal.pone.0069958

Seo, D. B., & Bernsen, M. (2016). Comparing attitudes toward e-government of non-users versus users in a rural and urban municipality. *Government Information Quarterly, 33*(2), 270–282. https://doi.org/10.1016/j.giq.2016.02.002

Shelton, T., & Lodato, T. (2019). Actually existing smart citizens. City, 23(1), 35–52. https://doi.org/10.1080/13604813.2019.1575115

Sieber, R. E., & Johnson, P. A. (2015). Civic open data at a crossroads: Dominant models and current challenges. *Government Information Quarterly, 32*(3), 308–315. https://doi.org/10.1016/j.giq.2015.05.003

Sieber, R. E., Robinson, P. J., Johnson, P. A., & Corbett, J. M. (2016). Doing public participation on the Geospatial Web. *Annals of the American Association of Geographers, 106*(5), 1030–1046. https://doi.org/10.1080/24694452.2016.1191325

Sweeting, D., de Alba-Ulloa, J., Pansera, M., & Marsh, A. (2022). Easier said than done? Involving citizens in the smart city. *Environment and Planning C: Politics and Space,* 23996544221080644. https://doi.org/10.1177/23996544221080643

Tulloch, D. L. (2008). Is VGI participation? From vernal pools to video games. *GeoJournal, 72*(3–4), 161–171. https://doi.org/10.1007/s10708-008-9185-1

Vicente, M. R., & Novo, A. (2014). An empirical analysis of e-participation. The role of social networks and e-government over citizens' online engagement. *Government Information Quarterly, 31*(3), 379–387. https://doi.org/10.1016/j.giq.2013.12.006

Wiseman, J. (2014). Can 311 call centers improve service delivery? Lessons from New York and Chicago (Issue Innovations in Public Service Delivery Issue No. 1).

Wu, W.-N. (2020). Determinants of citizen-generated data in a smart city: Analysis of 311 system user behavior. *Sustainable Cities and Society, 59,* Article 102167. https://doi.org/10.1016/j.scs.2020.102167

Zhang, S. (2019). Public participation in the Geoweb era: Defining a typology for geo- participation in local governments. *Cities, 85,* 38–50. https://doi.org/10.1016/j.cities.2018.12.004

Zhang, S., & Feick, R. (2016). Understanding public opinions from geosocial media. *ISPRS International Journal of Geo-Information, 5*(6), 74. https://doi.org/10.3390/ijgi5060074

Zook, M., Graham, M., Shelton, T., & Gorman, S. (2010). Volunteered geographic information and crowdsourcing disaster relief: A case study of the Haitian earthquake. *World Medical & Health Policy, 2*(2), 7–33.

Chapter 6
Passive Participation: Understanding Public Participation from Geosocial Media

Abstract Increasingly, social media data are linked to locations through embedded GPS coordinates. Many local governments are showing interest in the potential to repurpose these firsthand geo-data to gauge spatial and temporal dynamics of public opinions in ways that complement information collected through traditional public engagement methods. Using these geosocial data is not without challenges since they are usually unstructured, vary in quality, and often require considerable effort to extract information that is relevant to local governments' needs from large data volumes. Understanding local relevance requires development of both data processing methods and their use in empirical studies. This chapter addresses this latter need through a case study that demonstrates how spatially-referenced Twitter data can shed light on citizens' transportation and planning concerns. A web-based toolkit that integrates text processing methods is used to model Twitter data collected for the Region of Waterloo (Ontario, Canada) between March 2014 and July 2015 and assess citizens' concerns related to the planning and construction of a new light rail transit line. The study suggests that geosocial media can help identify geographies of public perceptions concerning public facilities and services and have potential to complement other methods of gauging public sentiment.

6.1 Introduction

Engaging citizens and other stakeholders is considered as an essential step in government decision-making (Innes & Booher, 2004). While public input has been collected traditionally through in-person techniques such as public meetings, workshops, and interviews, computer-aided technology has been used to supplement traditional methods (Brown, 2015; Cinderby, 2010). More recent developments in Web 2.0 and mobile technologies have drawn attention from public agencies and research communities seeking easier and less expensive methods of citizen engagement (Lin, 2013). Increasingly, local governments have been using social media platforms as additional communication channels to publish news and interact with citizens (Reddick & Norris, 2013). In addition, social media often contain information about

S. Zhang, *Geo-participation*, https://doi.org/10.1007/978-981-95-8228-0_6

public opinions and perceptions that is comparable to public comments collected through traditional public participation approaches (Massa & Campagna, 2014). It may potentially become a more convenient form of public participation as people are able to contribute information at any time from any location (Massa, 2015; Wukich & Mergel, 2016). Geotagged social media, also referred to as geosocial media (Croitoru et al., 2014; Kim et al., 2016), contain both descriptive comments and location information and thus may assist in understanding what the public needs are and where solutions need to be developed (Dunkel, 2015; Gal-Tzur et al., 2014a). However, unlike citizen surveys or interviews, social media data are the outputs of users' communication. Hence, these data are unstructured, vary in quality, and often of unknown relevance to local governments' need (Campagna et al., 2015). Further complications arise from the fact that only a small portion of social media are tagged with explicit geographic coordinates and that these data vary widely in their geographic representativeness within and across urban areas (Robertson & Feick, 2016). Their effectiveness in supporting public participation thus needs to be examined critically through empirical studies (Massa & Campagna, 2014).

This chapter aims to empirically examine the usability of geosocial media for local governments through a case study carried out in the Region of Waterloo (Ontario, Canada). We modeled the text content of geosocial media to identify commonly expressed topics and explored the spatial patterns of these identified concerns and interests. We believe that the insights drawn from the case study have the value in advancing our understanding of potential opportunities and challenges of using geosocial media for citizen participation. To facilitate the empirical study, a web-based Text Filtering and Analysis (TFA) toolkit that integrates several text analysis methods into an easy-to-use package was developed to ease the technical challenges of filtering irrelevant information from geosocial media and analyzing text content (Grant-Muller et al., 2015).

6.2 Literature Review

6.2.1 Use of Geosocial Media in Local Governments

Public participation is recognized as important since it can aid transparency and accountability in government and empower citizens in decision-making processes (Carver et al., 2001; Innes & Booher, 2004; Rowe, 2005). However, public participation has also been recognized as being a complex and contested process (Day, 1997; Higgs et al., 2008; Lowndes et al., 2001a). In particular, concerns have been centered upon issues including marginalized groups, effectiveness of participation approaches, and to what extent citizens are empowered in the participation process (Rhina Ghose, 2001; Rina Ghose, 2005; Irvin & Stansbury, 2004; King et al., 1998). The introduction of computer-based systems such as public participation GIS (PPGIS) was intended to address some of these challenges by providing integrated platforms for informing, creating, and sharing spatial knowledge (Hall et al., 2010; Sieber, 2006). Considerable effort has been made to develop mapping and visualization techniques that facilitate collecting and contextualizing spatial knowledge, identifying ways of enhancing collaborations among stakeholders, and engaging with marginalized populations using digital and Internet tools (Brown et al., 2014; Pocewicz et al., 2012). In some instances, however, these systems are criticized for their over-reliance on technical skills and high cost for development and maintenance (Slotterback, 2011).

Social media have been increasingly used by local governments in recent years because they provide an easy and inexpensive method of communication, and expand social networks through which governments can potentially reach large numbers of citizens (Evans-Cowley & Griffin, 2011; Reddick & Norris, 2013; Vicente & Novo, 2014). Although some governments use social media primarily to publish news and information, there is a trend of governments interacting with citizens through social media (Khan et al., 2013). According to Johannessen, Flak, and Sæbø (2012), local governments ranked social media in terms of preferred communication methods after email and websites. A growing body of literature has further examined how sentiments expressed in social media can help improve communication between local governments and citizens (Bonsón et al., 2015; Mossberger et al., 2013; Panagiotopoulos et al., 2014; Sandoval-Almazan & Ramon Gil-Garcia, 2014). Zavattaro, French, and Mohanty (2015), for example, suggested that social media sentiment is an effective indicator of successful interaction between local governments and citizens. Schweitzer (2014), similarly, identified strategies for transportation agencies to enhance their communication with the public through the analysis of transport-planning-related tweets.

In addition to their communication functions, social media are also considered as platforms for recording lived experiences of their users (Silva et al., 2013). That is, the way people tag places and events, check-in at venues, and comment leaves digital traces of their physical activities and reflects their personal opinions and sentiments (Li & Goodchild, 2012; Thelwall et al., 2012). Another line of studies has thus focused on mining geosocial media data that are spontaneously contributed by users. Several spatial analysis and visualization methods have been developed to derive public perception toward local environment and planning issues from social media. Dunkel (2015), for example, developed a visualization tool to help planners explore the perceived environment from geolocated Flickr data. Feick and Robertson (2015) proposed a multi-scale approach to identify commonality in how people define and delimit urban places in geotagged photo tags. Successful empirical studies of using geosocial media data for gauging public perceptions of disaster response, identifying events, and investigating human activities have also been documented (de Albuquerque et al., 2015; Panteras et al., 2014; Yin et al., 2012). In addition to the primary focus of mining spatial patterns, others suggested the need to further incorporate qualitative social media content (Campagna et al., 2015). Afzalan and Muller (2014), for example, found that informative dialogues are developed within online groups regarding local green infrastructure planning issues. Gal-Tzur et al., (2014a), similarly, suggested that useful information about transportation policy can be harvested from social media. The combination of qualitative and spatiotemporal analysis may, as suggested by Campagna et al. (2015), provide more insights into the geographies of public needs.

6.2.2 Challenges of Utilizing Geosocial Media

Incorporating qualitative analysis of geosocial media content presents several challenges. As argued by Afzalan and Evans-Cowley (2015), the relatively large volume of social media data may increase the time and human costs of analyzing the data and thus make the new data source less valuable. As a result, a number of scholars have explored the use of computer-aided methods to harvest and analyze information related to local government decision-making from social media (Campagna et al., 2015; Schweitzer, 2014). Gal-Tzur et al., (2014a), for example, suggested that text analysis methods can improve the efficiency of harvesting transportation-planning-related information from social media text. Campagna et al. (2015) similarly integrated basic text analysis functions such as generating tag clouds with spatiotemporal analysis to explore the use of location-based social media for spatial planning.

Notwithstanding these efforts, several challenges related to harvesting and analyzing locally specific social media remain (Massa, 2015). First, ontology-based information retrieval (IR) methods, which use concepts and their corresponding relationships to define domain-specific terminology and to recognize relevant text content, are used frequently to determine the relevance of a text message to a topic (Wang & Stewart, 2015). However, they are not entirely suitable for identifying locally relevant geosocial media messages, because: (1) there are few universal ontologies available for local government or even more defined fields such as planning (El-Diraby & Osman, 2011) and (2) many topics are location-specific and center on content that is relevant for a particular development plan or community. Although developing an ontology based on local knowledge is possible, such an ontology will be limited to a specific local context. Other commonly used IR methods such as machine-learning approaches have similarly been criticized for not being generic because of their need for large and good-quality training datasets (Kergosien et al., 2014). Second, individual social media messages need to be aggregated so that major needs or concerns can be identified and be further used for decision-making (Grant-Muller et al., 2015). Despite the growing use of computational methods, such as topic modeling methods, to automate interpretation of text data (Adams & McKenzie, 2013; Dou, Wang, Skau, Ribarsky, & Zhou, 2012; Ríos & Muñoz, 2014), manual work is still often employed to understand and categorize public input (Afzalan & Muller, 2014; Schweitzer, 2014). Moreover, the reliance on computational knowledge to use these analytical methods can be a barrier to government adoption of social media as it may increase both financial and human cost (Johnson & Sieber, 2013).

Additionally, there are concerns about whether geosocial media data can meet local governments' needs. Because social media are networking and communication platforms at the core, their users are mostly contributing information without being aware that it might be used for other purposes (Stefanidis et al., 2013). Although this may result in less guarded recordings of user sentiment (Dunkel, 2015), much of the data may be irrelevant to local government needs. Further empirical studies of the nature and value of information that can be harvested from social media are therefore needed (Gal-Tzur et al., 2014a). Moreover, the geographic representativeness of geosocial media varies widely both among different cities and within cities (Robertson & Feick, 2016). Several studies have reported the geographical unevenness of volunteered geographic information, in that some areas may be represented by large amounts of data, while other areas have very few (Graham et al., 2015; Hardy, 2013; Hollenstein & Purves, 2010). Varying demographic profiles across geographic areas may also contribute to the geographical unevenness of user-generated content. As noted by Cavallo et al. (2014), certain population groups may opt out of using new digital technologies. Although this unevenness should not mitigate the value of user-generated geographic information, the use of these data need to be critically examined through context-specific analysis (Cavallo et al., 2014).

In this regard, local government adoption of geosocial media needs solutions to alleviate technical challenges of utilizing the data as well as further recognition of the local relevance and potential limitations of the data through critical examinations.

6.3 Methodology

Based upon previous sections, three components are essential for local government staff to harvest and analyze relevant information from social media (Fig. 6.1). A web-based toolkit was developed to: (1) harvest geosocial media data from online sources; (2) identify text-based geosocial media messages that relate to local spatial planning issues; and (3) semi-automatically summarize the text content and explore main themes that appear from public input. These themes are delivered through an interactive visual design to help local authorities understand the data.

6.3.1 Data Collection

We chose Twitter as an exemplar social media service as it is one of the most popular microblogging services for users to post text messages, share images, tag locations, and interact with others. Twitter has a large user base—one in every ten American adults gets news from Twitter (Schweitzer, 2014), and while its user community is skewed toward affluent and educated individuals, it is reportedly more diverse than is found on other social media platforms (Lenhart & Fox, 2009). Twitter data were collected using Twitter's Application Programming Interfaces (APIs) and an associated Python library Tweepy (http://www.tweepy.org/). Only data that contain valid geographic coordinates were collected for a local study area and were stored in PostgreSQL database after parsing time, spatial, and user information. For other

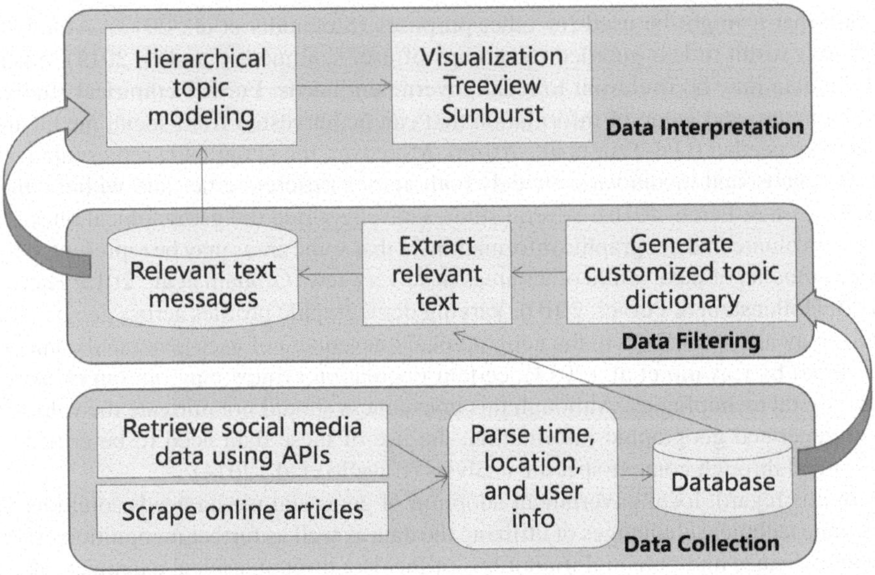

Fig. 6.1 The workflow of collecting and analyzing social media data

sources of user-generated content such as online articles and citizen letters, Python scripts built with the scrapy library (http://scrapy.org/) were used to extract information directly from web pages. These text documents were also stored in the database with ancillary information such as the source and time.

6.3.2 *Extraction of Relevant Geosocial Media Text Messages*

A two-step approach is used to identify social media messages that relate to local topics (Fig. 6.2). As discussed in Sect. 6.2.2, one challenge associated with extracting text messages relevant to a local planning context is the need of locally specific resources (e.g., ontology, training datasets, etc.). A more generic approach is used here to build a local lexicon from local news, municipal reports, and articles based on the widely used tf-idf metric. This lexicon is then used as input to evaluate the relevance of the text messages based on a language-modeling approach that is found to be effective for identifying relevant short text messages from social media (Tao et al., 2012).

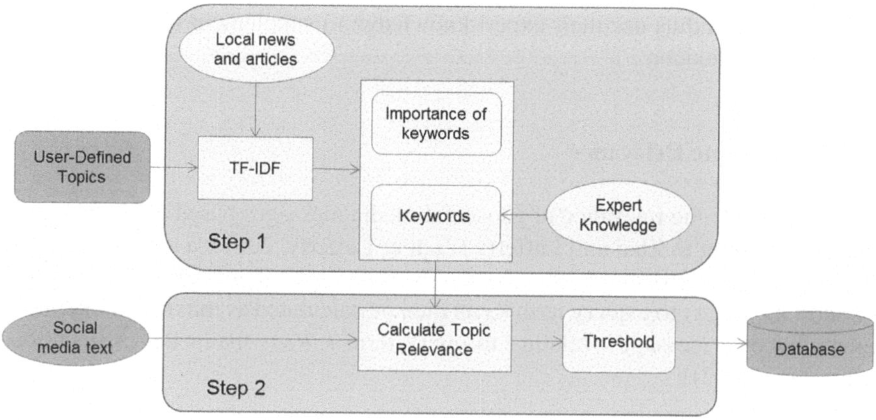

Fig. 6.2 A two-step procedure to automatically identify relevant social media messages

Constructing Local Lexicon

A local lexicon composed of domain- and context-specific terms is built based on news postings, government documents, and articles that relate to a topic or issue of interest to a local government (e.g., public transportation, infrastructure, construction, etc.). In particular, the tf-idf measurement is applied to identify the most important words from collected articles. The method considers both the occurrence of a word in a document and the uniqueness of a word according to the number of documents it occurs within so that it can reduce the effect of common words, which are words that generally occur more often than others in a language (Rattenbury & Naaman, 2009). For each word in the corpus, a tf-idf value is calculated using Eq. (6.1):

$$tf - idf = \frac{count(w, d)}{size(d)} \times \log\left(\frac{n}{docs(w, D)}\right), \tag{6.1}$$

where a term frequency (tf) is first calculated using the number of times a word w occurs in a document d (count (w, d)) and the total number of words the document d contains (size(d)). This tf value is then multiplied by an inverse document frequency (idf) value, which is an inverse fraction of the total number of documents n and the number of documents that contain word w (docs (w, D)), to get the tf-idf value for the word w. The higher the tf-idf score is, the more important the word w is. A list of important words with high tf-idf scores is then used to generate a customized local lexicon based on the assumption that important words identified from planning-related documents are more likely to be related to planning topics (Gal-Tzur et al., 2014a). In addition to keywords derived from local documents (e.g., "parking"), their semantic variants (e.g., "parking lot," "parked," "parking garage") can be included in the lexicon to improve the accuracy of the IR (Abel et al., 2011). Government professionals can thus use their expert knowledge to supplement or alter the auto-generated local lexicon.

Calculating Topic Relevance

We then evaluate the relevance of geosocial media messages based on the language model. According to Zhai and Lafferty (Zhai & Lafferty, 2001), a text message can be considered as a probability distribution over the words it contains. The relevance of a short message t to a query term k can then be calculated as maximum likelihood probability of a message t relating to query term k $R_{(t|k)}$ using Bayes likelihood estimate (Eq. (6.2)):

$$R_{(t|k)} = \frac{C(k, t) + uP(k|\theta_L)}{len(t) + u} \tag{6.2}$$

In the equation, $C(k, t)$ is the times the query k occurs in the message t; P is the probability of the query term k occurs in the whole corpus; $len(t)$ is the length of the message t; u is a smoothing parameter for Bayes estimate.[1]

To evaluate the relevance of a text message to a topic, we consider each topic as a collection of query terms, which correspond to the keyword and its semantic variants as derived from the previous step. Therefore, a topic T can be represented as: T = $\langle k_1, k_2, k_3, \ldots, k_n \rangle$, where n is the total number of keywords identified for topic T. The relevance of a text message to a topic can then be evaluated using Eq. (6.3):

$$R_{(t|T)} = \sum_{i=0}^{n} \frac{W(k_I|T)C(k_i, t) + uP(k_i|\theta_L)}{len(t) + u} \tag{6.3}$$

Here, the relevance of a message to a topic is considered as a sum of the message's relevance to each word in the topic dictionary. Each word is weighted using its tf-idf score to decrease the effect of less important or common words in the dictionary. Using this method, each message will receive a relevance score indicating its relevance to a topic T, with a higher score suggesting a higher possibility of being relevant.

A threshold is then determined to differentiate relevant messages from irrelevant ones by reviewing a sample of messages and their corresponding scores. As shown in Table 6.1, although all the selected text messages refer to parking expressions, the first four with higher scores are potentially of more interest to planners, whereas the latter two relate more to personal feelings. A larger sample of Twitter tweets can be reviewed using the same method to determine the appropriate threshold for identifying parking-related text messages. Although somewhat subjective, reviewing

[1] After manually testing a series of smoothing parameters, u = 0.1 is used in the equation for the best accuracy of Bayes estimate.

Table 6.1 Evaluating the relevance of a message to topic "parking"

Text	Score	Category
"This mcdonalds parking lot is so unnecessarily massive."	0.247	Relevant
"why does Waterloo parking charge so much for parking lots that are OBVIOUSLY over capacity everyday. open rim parking lots for f*** sake"	0.173	Relevant
"The freshco parking lots so d*** complicated. I'm almost prone to get into an accident"	0.132	Relevant
"Feels like the entire population of Waterloo is in this parking lot. can't find a place to park"	0.105	Relevant
"@570News turkey on the loose at Canadian Tire Weber Street parking lot! http://t.co/rXlmjXQvOZ"	0.076	Irrelevant
"I don't know why, but I always feel more comfortable parking next to other BMWs. So I've turned it into a game. I search them out"	0.036	Irrelevant

a relatively small sample of the data allows local government staff to view more details about the data and bring expert knowledge to the categorization procedure.

6.3.3 Understanding Public Input Using Hierarchical Topic Modeling

Having identified relevant geosocial media messages, a topic modeling approach is used to recognize latent sub-topics within message collections. Topic modeling is a suite of text mining methods for identifying semantic patterns within collections of natural language documents (Blei, 2012). The Latent Dirichlet Allocation (LDA) was selected because of its simple yet powerful nature (Adams & Janowicz, 2015; Adams & McKenzie, 2013; Blei, 2012). Each topic is associated with a list of keywords, based on which meanings of topics can be interpreted.

Within the context of this work, geosocial media messages related to a topic T are considered as a corpus, which the LDA method divides into a collection of sub-corpora. Assume topic T is "cycling," the above method would allow us to identify what aspects (e.g., cycling trails, shared-use path, and safety concerns) of "cycling" people are talking about. The same procedure can be repeated for these sub-topics to reveal more details from the text. Python scripting is used to automate this recursive procedure following the logic as shown in Fig. 6.3. Topic models are first generated for the entire corpus. The words in the corpus are then reassigned to a set of new corpora based on their relationship with the topics. The procedure is repeated for each new corpus until the number of messages the corpus contains is less than a minimum threshold. As a result, texts are modeled as a topic hierarchy that is composed of various topic paths, which represent how one topic is broken down into several sub-topics.

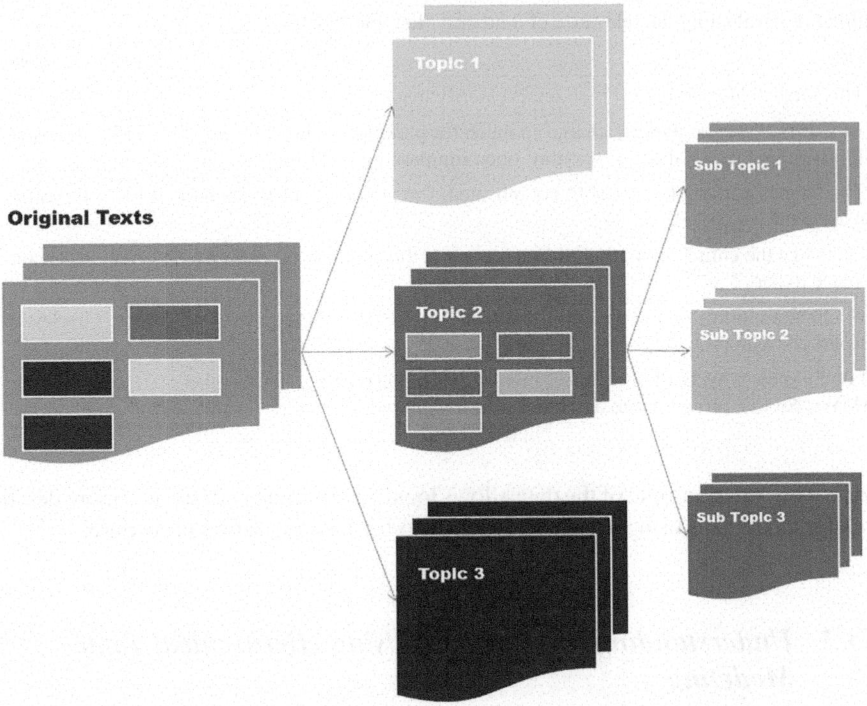

Fig. 6.3 The logic of hierarchical LDA (hLDA) (adapted from Smith et al., 2014)

6.3.4 Design and Implementation of a Web-Based Tool

The Django-based TFA toolkit was developed to provide an easy-to-use graphical interface that integrates IR and the topic modeling method. Django is a free and open-source framework for web development (Django, n.d.). Figure 6.4 shows the system architecture of the application. On the backend, a PostgreSQL database is used to store parsed text messages as well as spatial and temporal information. On the server side, a series of models that process and analyze text data is developed using python scripts built from the open-source natural language processing (NLP) python library NLTK. GeoDjango handles the reading and storage of spatial data including locations of social media messages. On the browser side, map visualizations are generated using Leaflet and topic modeling results are visualized with the popular JavaScript-based D3 visualization library (see the D3 Gallery—https://github.com/mbostock/d3/wiki/Gallery).

Figure 3.5 shows several screenshots of the toolkit. Users can follow the steps on the left panel of the main interface to harvest and analyze text input (Fig. 6.5a). Customized local lexicons can be created by selecting topic-related documents or by specifying online sources to scrape articles from (Fig. 6.5b). Amendments can then be made to the auto-generated keyword list for identifying relevant text

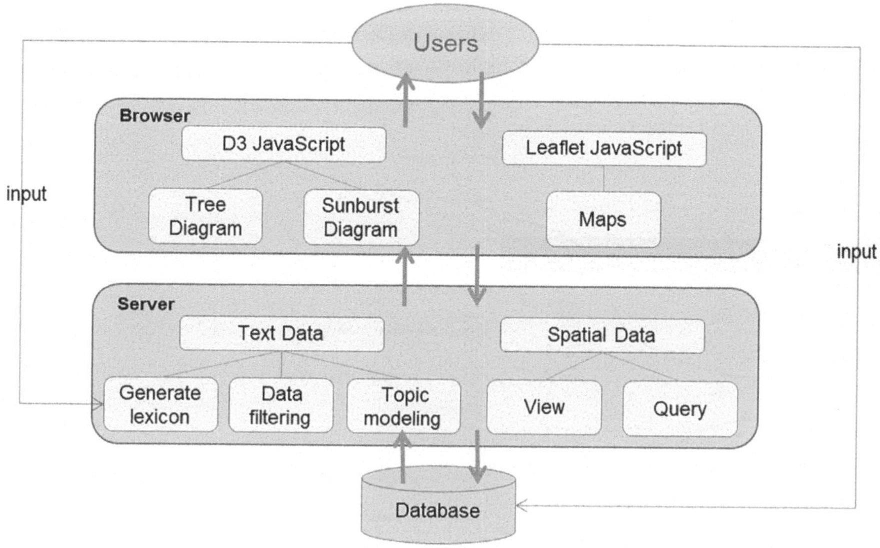

Fig. 6.4 System design of TFA toolkit

messages (Fig. 6.5c). Clusters of relevant tweets are represented on the map using the Leaflet markercluster library (https://github.com/Leaflet/Leaflet.markercluster). Topic modeling results are then displayed as shown in Sect. 3.4.

6.4 Case Study

To demonstrate the possible value of topic modeling and mapping of geosocial media data, the toolkit described above was applied in the cities of Waterloo and Kitchener within the Region of Waterloo, Canada (Fig. 6.6). The Region of Waterloo has consistently been ranked as one of the fastest growing communities in Canada and is forecast to increase in population from its current level of 568,500 to 729,000 by 2031 (Region of Waterloo, 2007). Consulting stakeholders is and will continue to be an important function for local governments as the development unfolds. The ongoing construction of a new light rail transit (LRT) started in August 2014, has promoted public debate concerning issues such as congestion, urban intensification, and disruptions to existing neighborhoods. During the preparation of the project, both the regional and the city governments held public meetings to collect public opinions toward the transit plan at different stages of the project. Discussions about the project are continuing as the impacts of the LRT construction and associated intensification of urban forms become more apparent to local residents.

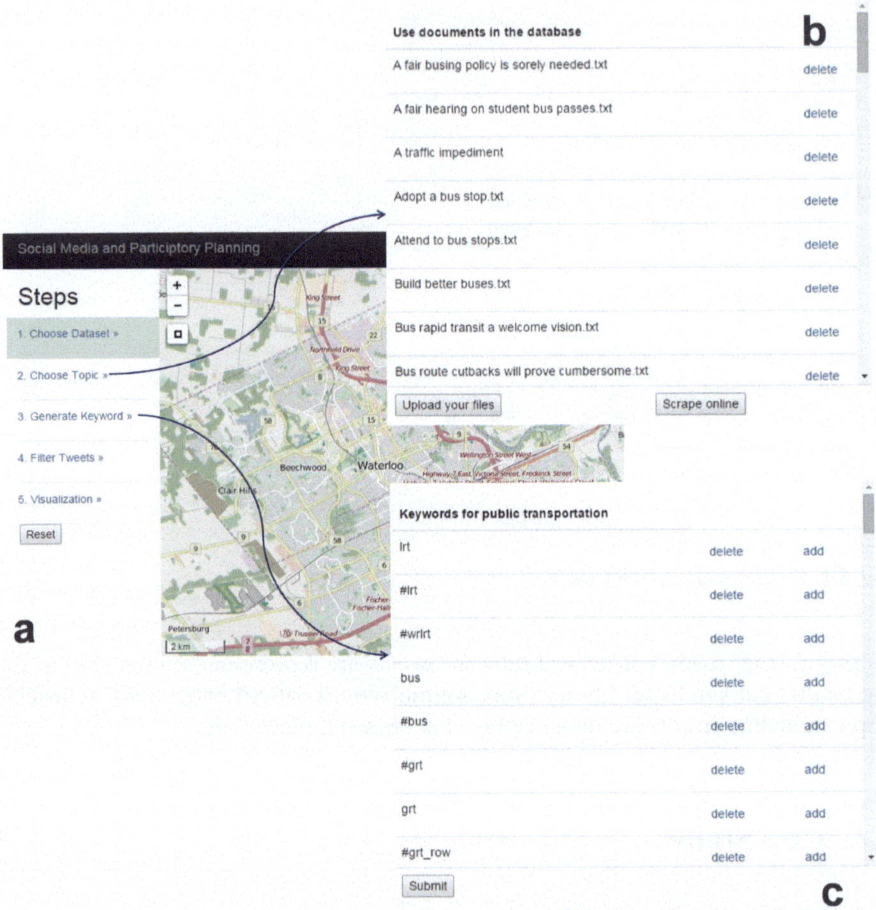

Fig. 6.5 (a) The main interface of the TFA toolkit; (b) selecting documents for generating a customized topic lexicon; and (c) reviewing and modifying the auto-generated keyword list

6.4.1 Data

Twitter data with valid geographic coordinates were obtained in real time from March 2014 to July 2015, the time period when the Region started constructing the first stage of LRT, based on a fixed boundary for cities of Waterloo and Kitchener. It is important to note that although only some one percent of tweets can be obtained using public streaming API, the absolute quantity of the sample is still relatively large (Leetaru et al., 2013). In the following analysis, we focus on transportation-related topics given the ongoing LRT project has elevated the issue of transportation within the Region and the general importance of transportation in many other locales (Gal-Tzur et al., 2014b).

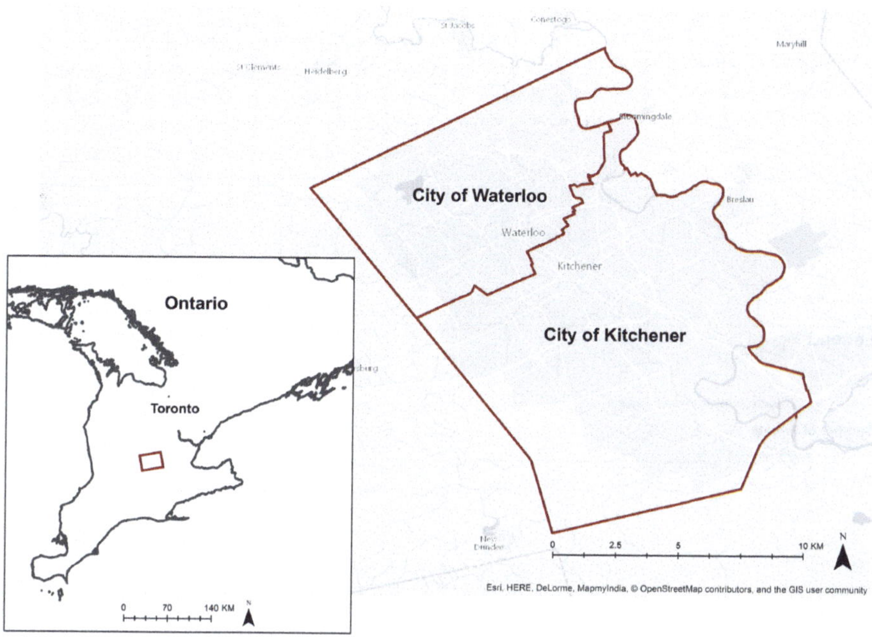

Fig. 6.6 Study area

A topic dictionary based on a purposely restrictive keyword set (LRT, light rail, bus, public transportation, GRT) was developed by scraping news and commentary articles from local media ("The Record" newspaper, http://www.therecord.com/waterlooregion/). Over 200,000 Twitter tweets with valid geographic coordinates were collected during the 16-month period. In total, 2777 and 2112 tweets were found to be relevant to the topics "public transportation" and "walking," respectively. This volume is similar to that found in de Albuquerque et al. (de Albuquerque et al., 2015), where over 99% of Twitter tweets were found to be "off-topic."

To test the accuracy of the results, we manually classified a random sample (sample size = 120) for each topic and compared the results with computer-coded ones. We found 82.5% and 67.5% precision respectively for public transportation and walking. Interestingly, some messages about the TV show "The Walking Dead" were mistakenly classified as walking-related because the term "walking" has the highest weight in the lexicon. To improve this result, we adjusted the weight of the word "walking" and the relevance threshold of the topic accordingly. Testing of another randomly generated sample suggested that the precision of classification results increased to 80.83%, which is reasonable for IR of short text messages (Tao et al., 2012). We further examined the spatial distributions of these messages to draw insights into their locational context (Figs. 6.7 and 6.8). The maps shown here were reproduced in ArcGIS to add map elements such as legends, scales, and better-quality graphs. In addition to mapping individual locations of messages, clustering circles

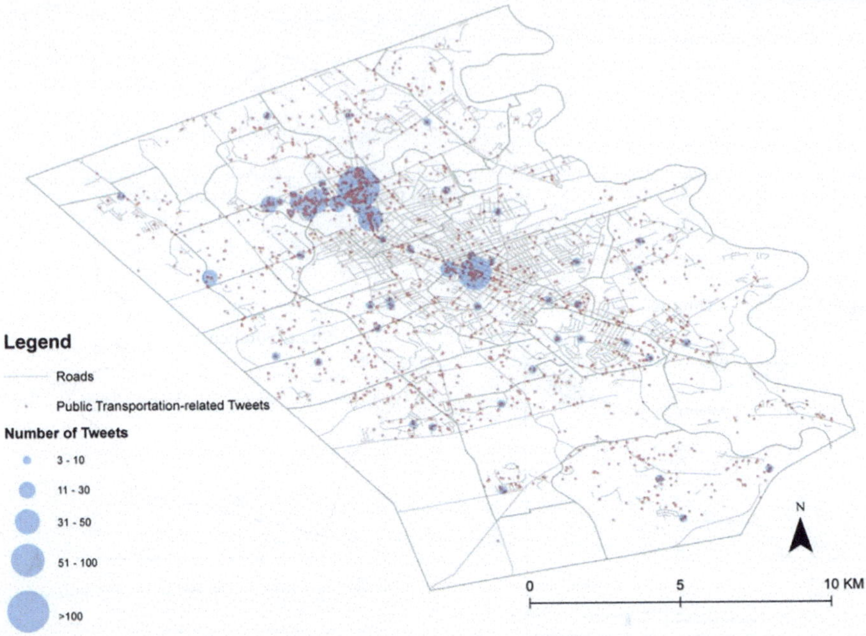

Fig. 6.7 Spatial distributions of public transportation-related tweets

are also mapped with the size indicating the counts of tweets within the area. As expected, most tweets were posted nearby two universities (University of Waterloo and Wilfrid Laurier University), and the cores of Waterloo and Kitchener, as those are the busiest areas where most students and business are located.

To further investigate the content of these messages, the topic modeling method was applied to find major topics of interest to Twitter users. The keyword list originally produced by LDA is shown in Table 6.2. While the relevance of some topics is evident, other terms shown in italics were less helpful and were removed from the topic hierarchy (Table 6.2).

6.4.2 *Understanding Public Perception from Geosocial Media*

Figure 6.9 shows a sunburst diagram generated based on topic modeling results of public transportation-related messages. Five topics, including trains, bus services, Uptown Waterloo, Charles Terminal (shortened from Charles Street Bus Terminal), and LRT, are found at the top of the hierarchy (the second inner-most ring in Fig. 6.9). Among these five topics, three are associated with public transportation modes

Fig. 6.8 Spatial distributions of walking-related tweets

Table 6.2 Examples of keyword list generated from LDA

Topic	Personal feelings	Long distance	Schools and kids	Schools and kids	Irrelevant	Irrelevant
Keyword	safety	walk	off	school	nice	nice
	comfort	far	school	walk	love	respect
	walking	school	kids	dont	im	like
	watch	away	snow	ice	get	dont
	like	campus	far	friendly	places	soccer
	lights	friendly	kitchener	back	home	people
	catch	minutes	traffic	ill	hate	S***
	dog	class	hate	lights	turning	got
	good	snow	mom	waterloo	guys	one
	time	car	ice	cold	tonight	weather

(trains, bus, and LRT), and the other two relate to two transit hub locations (Uptown Waterloo, Charles Terminal).

While these topics generally provide a high-level overview of public transportation messages, more details are revealed at the next levels of the hierarchy. For example, the topic "bus service," located in the lower left of Fig. 6.9, is split into "bus

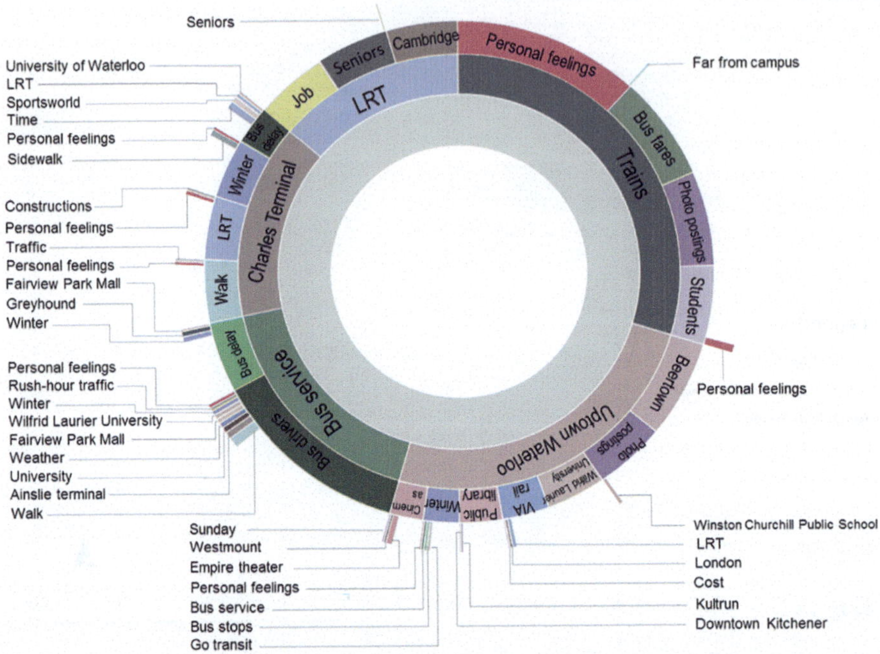

Fig. 6.9 An overview of topic hierarchy generated from Twitter Tweets

delay" and "bus drivers" topics. Given that a few studies have suggested that social media comments are more negative rather than positive (Schweitzer, 2014) and that "bus delay" itself is not a positive expression, it is reasonable to speculate that "bus delay" is the aspect that people have the most complaints about. In other instances, topics may occur multiple times in the hierarchy yet indicate different contexts. For example, "winter" appears under both "Charles Terminal" and "Uptown Waterloo" and relates to the infrastructure. Near Charles Terminal, people mentioned concerns with sidewalks in the surrounding area, largely because of ongoing construction (e.g., "*@CityKitchener can you please fix sidewalk bricks queen at King to Charles. I'm tired of twisting my ankles on missing bricks.*"). In Uptown Waterloo, "winter" was used more frequently to register a complaint about the lack of shelters at some bus stops (e.g., "*we need a shelter at bus stop #1908*"). This type of information is typical of what is reported through various open 311 applications that permit citizens to report concerns with city infrastructure and public services (Sieber & Johnson, 2015).

Text analysis results can also be combined with geolocations to examine where concerns or interests are expressed. Figure 6.10 shows major locations where tweets related to the topic "bus service" are posted. Spatial clusters were mapped using the proportion of bus service-related tweets to the total amounts of public transportation tweets within the same location in order to mitigate the effect of varying numbers of tweets in different locations. Not surprisingly, messages under this category are

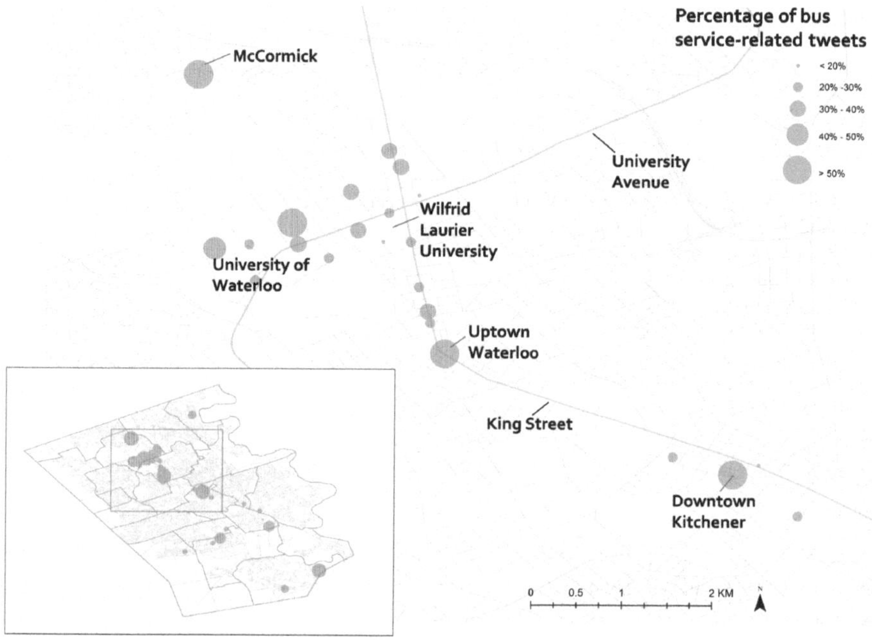

Fig. 6.10 Spatial clustering of bus service-related tweets

mostly concentrated around the University area as well as King Street, the central transit corridor in the Region. Several residential areas with concentrated rental housing in the northern Waterloo and southern Kitchener also appear to be significant, indicating high usages of bus service. General insights can be drawn from the map on where common concerns and needs are. For example, messages about bus drivers are consistently seen around Downtown Kitchener, indicating some pertinent traffic issues such as narrow road lanes and busy traffic in the area.

Other locations may not appear to be significant using the data collected for the entire sixteen months but become more visible within certain time periods. For example, Fig. 6.11 illustrates how road closures and changes in bus routes in June 2015 are reflected in bus-service-related Tweets before and after June 2015. These changes were required at this time to permit a new LRT station to be constructed near the two tweet clusters close to the Parkside/Northfield intersection. While this is a specific example, it provides some indications of how geosocial media may help identify the dynamics of public opinions. Local governments can potentially use this data to examine the effects of planning and development projects on local people in a timelier manner.

Yet, public opinions expressed through geosocial media mostly relate to public sentiment and perceptions toward their immediate environment. For example, although there is a growing trend of LRT-related tweets (from an average of 8% of all public transportation-related messages in 2014 to an average of 12% in 2015)

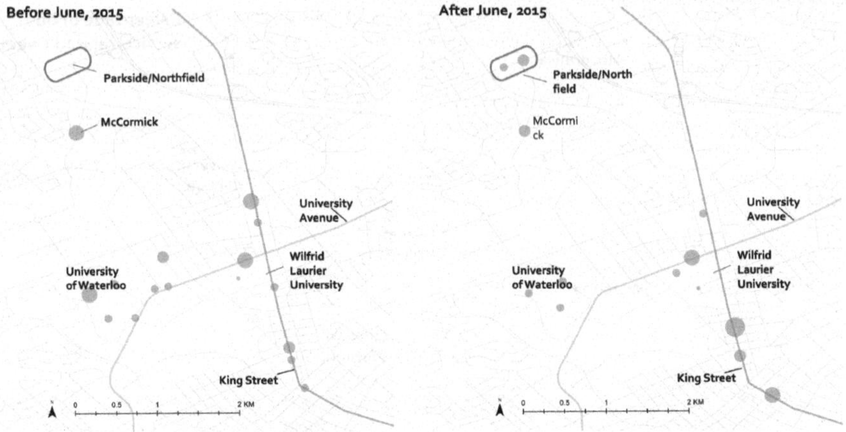

Fig. 6.11 The comparisons of bus service-related tweets before and after June 2015

because of the ongoing construction, the discussion of LRT mainly reflects users' experience with current traffic situation (e.g., *"On a jam-packed express bus—a good harbinger of ridership for the ION light rail line! Another reminder of how excited I am."*). This also indicates the difference between geosocial media and traditional participation methods, which will be further examined in the next section.

6.4.3 Comparing Different Forms of Citizen Input

In addition to geosocial media, many traditional public participation methods, such as open house events, workshops, citizen letters, and surveys, also collect text and often geographically referenced input from citizens. This input can be analyzed in the same way as what was done for geosocial media messages. On a regular basis, The Record publishes citizen letters and comments that relate to public concerns. In total, 478 transportation-related citizen letters were obtained during the same time period as Twitter data were collected. These letters were processed following the same procedure as shown in Fig. 6.3. Figure 6.12 shows an overview of topics that emerged from the content of citizen letters.

An initial examination of topic categories demonstrates marked differences between geosocial media messages and citizen letters. Some topics, such as roundabout, traffic lights, and disabled passengers, do not occur in social media messages, whereas social media messages have other unique topics which mostly are place-based (e.g., Fairview Park Mall, Ainslie Terminal, and Beertown—a restaurant) and event-oriented. Although both citizen letters and Twitter messages mention certain places, places mentioned in citizen letters more refer to general areas, such as school zones, university, etc., whereas more specific place names are mentioned in Twitter messages. This general versus specific distinction is comparable to what others found

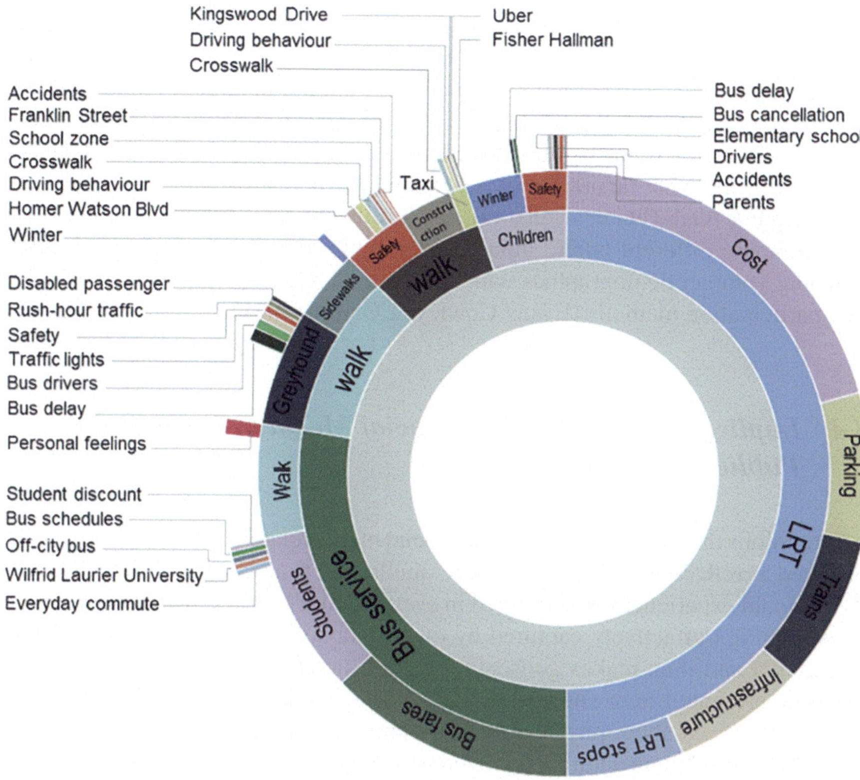

Fig. 6.12 An overview of topic hierarchy generated from citizen letters

in comparison of walking and sedentary interviews (Evans & Jones, 2011). Similar to walking interviews, Twitter messages can better capture the dynamics of urban landscape as people often send messages when they are moving around the city. Citizen letters, analogous to sedentary interviews, serve as a more productive mode for narratives and an incubator of critical and deeper discussions on issues such as safety, urban design, and policy. Another unique characteristic of Twitter, or social media in general, is its capability to capture events and activities. In the topic hierarchy generated from Twitter messages, photo-posting activities appear to be associated with Uptown Waterloo and trains. Many messages in this category relate to an "IONUptown" challenge that was launched by Uptown Waterloo Business Improvement Area (BIA) office (http://uptownwaterloobia.com/ionuptown-challenge/#). Many people, incentivized by the possibility of winning a prize, were willing to participate in the challenge by posting photos on Twitter about their work, play, or shopping activities around the Uptown area using hashtag #ionuptown and had a chance to win a prize. Methods demonstrated by Dunkel (2015) to examine the photo content in addition to the text tags are beyond the scope of this study, but could be used in future analysis to learn more about citizens' place perceptions and preferences.

Moreover, even topics that occur in both datasets may have completely different foci. While social media users mostly talked about bus delays and bus drivers regarding bus service, citizen letters demonstrate a quite divergent range of issues related to students, walking, and costs. These differences most likely can be traced to the different nature of the two input methods, one more temporally immediate and place-specific, the other favoring more contemplative and geographically generic, as discussed above (Gal-Tzur et al., 2014a). On the other hand, it provides an interesting lens to compare different public input, especially on the potential of social media in reaching younger demographics which are often under-represented in traditional public participation methods (Evans-Cowley & Griffin, 2011).

6.4.4 *Implications for Using Geosocial Media to Understand Public Opinions*

Public opinions that can be retrieved from social media relate to what Corburn (2003) considered as reflections of "actual sights, smells, and tastes, along with the tactil(e) and emotional experiences encountered in everyday life" (P.421). Knowledge of this kind is often not effectively captured by other data collection methods (McCall & Dunn, 2012) and thus makes geosocial media a potentially valuable source. The case study presented here suggests that geosocial media can help identify public concerns and needs about physical facilities and the quality of public services, and potentially be used as an additional citizen reporting mechanism. Moreover, as illustrated in the case study, messages about the LRT project appear shortly after the start of the construction, suggesting a potential use of geosocial media to capture the dynamics of public perception over space and time. In addition, public perception expressed through Twitter is often a reflection of people sensing and responding to their immediate environments and differs from public input collected from formal public participation procedures, which is usually given based on more considered thought and rational choice (Gal-Tzur et al., 2014a).

With regard to spatial bias in geosocial media, the uneven geographic distributions of tweets were not surprisingly found to be concentrated within university areas, city-core areas, and the major transit corridor, while data points in other areas were relatively sparse. As suggested in other studies, this unevenness may limit the use of geosocial media to certain areas (Lawrence et al., 2015; Shelton et al., 2015). However, we were able to identify places outside high-interaction areas that were associated with particular topics or emerged at specific time periods. To understand spatial bias in data of this type, some attention should be directed to exploring qualitative analysis at different spatial and temporal scales.

The comparison between geosocial media and citizen letters further investigates the differences between geosocial media and other methods of monitoring public sentiment. Although geosocial media may be limited in providing more in-depth discussion and comments in response to local government initiatives, they illustrate some potential of complementing other public participation methods as well as fostering new virtual interactions between government and citizens through online activities. These findings have several implications for citizen–government interactions. First, geosocial media may assist the study of "the relationship between what people say and where they say it" (Evans & Jones, 2011), which is found to be a challenging task because of the difficulty in identifying locational information from interviews (Elwood & Martin, 2000). While people are found to mention general areas more often in formally written comments such as citizen letters, geosocial media could supplement other methods by identifying where certain issues may be worth further exploration. Second, the response to the IONUptown challenge suggests that there is a good potential to boost citizen contributions through entertaining place-based activities.

However, local government professionals' perspectives will be critical for evaluating these identified possibilities and challenges. In practice, government adoption of social media as a monitoring mechanism depends not only on whether valuable information can be identified from social media, but also on various factors such as the trustworthiness of data contributors and the organization's culture with respect to adapting to new technologies (Grant-Muller et al., 2015; Wukich & Mergel, 2016). Future work will examine the case-study findings further by interviewing local government professionals.

6.5 Conclusions

This chapter was intended to address challenges of utilizing geosocial media and assess the potential of these data sources as a new channel for gathering place-based public opinions. The potential uses and challenges identified from the case study contribute to an emerging body of literature on local governments' adoption of social media. The empirical study illustrates how geosocial media can provide topic- and location-specific types of public input that differ subtly from what might be found in complementary data sources. Second, based on the inevitable geographic unevenness of geosocial media data, our study suggests that such an unevenness should be explored further by incorporating qualitative analysis at different spatial and temporal scales. Additionally, different from many geosocial media studies focusing on metropolitan cities, we purposefully chose cities of Waterloo and Kitchener to shed light on whether perceived opportunities of geosocial media are applicable to medium-sized cities. Finally, the TFA toolkit facilitated the study by alleviating technical challenges for harvesting and analyzing social media content. Designed for social media messages, this toolkit can be used for other text-based public input, such as that collected from surveys, public meetings, online forums, and different social media platforms. Further user study is needed to test the functionality and user-friendliness of the toolkit in order to broaden its usage.

There are several ways where the use of geosocial media in a local government context can be further explored. First, our analysis focuses on Twitter, which is only one of the most popular social media platforms. It will be worthwhile to examine whether an integration of various types of social media would allow different subpopulations to be represented and different aspects of behavior and interaction to be captured. Second, a relatively small proportion of social media have encoded geographic coordinates. Georeferencing implicit spatial information such as place names may enrich data volume and increase the potential to glean useful information from social media. Finally, future work should further combine spatiotemporal analysis with ancillary information such as user profiles to uncover the representativeness of geosocial media and advance our understanding of how geosocial media may complement other participation methods.

References

Abel, F.; Celik, I.; Houben, G. J.; Siehndel, P. Leveraging the semantics of tweets for adaptive faceted search on twitter. Lect. Notes Comput. Sci. (including Subser. Lect. Notes Artif. Intell. Lect. Notes Bioinformatics) 2011, 7031 LNCS, 1–17.

Adams, B., & McKenzie, G. (2013). Inferring thematic places from spatially referenced natural language descriptions. In D. Z. Sui, S. Elwood, & M. F. Goodchild (Eds.), *Crowdsourcing geographic knowledge: Volunteered geographic information (VGI) in theory and practice* (pp. 201–221). Springer.

Adams, B., & Janowicz, K. (2015). Thematic signatures for cleansing and enriching place-related linked data. *International Journal of Geographical Information Science, 29,* 556–579.

Afzalan, N., & Muller, B. (2014). The role of social media in green infrastructure planning: A case study of neighborhood participation in park siting. *Journal of Urban Technology, 21,* 67–83.

Afzalan, N., & Evans-Cowley, J. (2015). Planning and social media: Facebook for planning at the neighbourhood scale. *Planning Practice & Research, 30*(1), 11–26.

Blei, D. M. (2012). Introduction to probabilistic topic modeling. *Communications of the ACM, 55,* 77–84.

Bonsón, E., Royo, S., & Ratkai, M. (2015). Citizens' engagement on local governments' Facebook sites. An empirical analysis: The impact of different media and content types in Western Europe. *Government Information Quarterly, 32*(1), 52–62.

Brown, G., Weber, D., & de Bie, K. (2014). Assessing the value of public lands using public participation GIS (PPGIS) and social landscape metrics. *Applied Geography, 53,* 77–89.

Brown, G. (2015). Engaging the wisdom of crowds and public judgement for land use planning using public participation geographic information systems. *Australian Planner, 52,* 199–209.

Campagna, M., Floris, R., & Massa, P. (2015). The role of social media geographic information (SMGI) in spatial planning. In S. Geertman, J. Stillwell, & R. Ferreira (Eds.), *Planning support systems and smart cities* (pp. 41–60). Springer.

Carver, S., Evans, A., Kingston, R., & Turton, I. (2001). Public participation, GIS, and cyberdemocracy: Evaluating on-line spatial decision support systems. *Environment and Planning b: Planning and Design, 28,* 907–921.

Cavallo, S., Lynch, J., & Scull, P. (2014). The digital divide in citizen-initiated government contacts: A GIS approach. *Journal of Urban Technology, 21,* 77–93.

Cinderby, S. (2010). How to reach the "hard-to-reach": The development of Participatory Geographic Information Systems (P-GIS) for inclusive urban design in UK cities. *Area, 42,* 239–251.

Corburn, J. (2003). Bringing local knowledge into environmental decision making. *Journal of Planning Education and Research, 22,* 420–433.

Croitoru, A., Wayant, N., Crooks, A., Radzikowski, J., & Stefanidis, A. (2015). Linking cyber and physical spaces through community detection and clustering in social media feeds. *Computers, Environment and Urban Systems, 53,* 47–63.

Day, D. (1997). Citizen participation in the planning process: An essentially contested concept? *Journal of Planning Literature, 11,* 421–434.

de Albuquerque, J. P., Herfort, B., Brenning, A., & Zipf, A. (2015). A geographic approach for combining social media and authoritative data towards identifying useful information for disaster management. *International Journal of Geographical Information Science, 29,* 667–689.

Django The Overview of Django Framework. Retrieved January 9, 2016, from https://www.djangoproject.com/start/overview/

Dou, W., Wang, X., Skau, D., Ribarsky, W., & Zhou, M. X. (2012). LeadLine: Interactive visual analysis of text data through event identification and exploration. In *IEEE Conference on Visual Analytics Science and Technology (VAST)* (pp. 93–102). IEEE.

Dunkel, A. (2015). Visualizing the perceived environment using crowdsourced photo geodata. *Landscape and Urban Planning, 142,* 173–186.

El-Diraby, T. E., & Osman, H. (2011). A domain ontology for construction concepts in urban infrastructure products. *Automation in Construction, 20,* 1120–1132.

Elwood, S. A., & Martin, D. G. (2000). "Placing" interviews: Location and scales of power in qualitative research. *The Professional Geographer, 52,* 649–657.

Evans, J., & Jones, P. (2011). The walking interview: Methodology, mobility and place. *Applied Geography, 31,* 849–858.

Evans-Cowley, J. S., & Griffin, G. Micro-participation: The role of microblogging in planning. *Social Science Research Network Electronic Journal, 2011,* 1–38.

Feick, R., & Robertson, C. (2015). A multi-scale approach to exploring urban places in geotagged photographs. *Computers, Environment and Urban Systems, 53,* 96–109.

Gal-Tzur, A., Grant-Muller, S. M., Kuflik, T., Minkov, E., Nocera, S., & Shoor, I. (2014a). The potential of social media in delivering transport policy goals. *Transport Policy, 32,* 115–123.

Gal-Tzur, A., Grant-Muller, S. M., Minkov, E., & Nocera, S. (2014b). The impact of social media usage on transport policy: Issues, challenges and recommendations. *Procedia—Social and Behavioral Sciences, 111,* 937–946.

Ghose, R. (2001). Use of information technology for community empowerment: Transforming geographic information systems into community information systems. *Transactions in GIS, 5,* 141–163.

Ghose, R. (2005). The complexities of citizen participation through collaborative governance. *Space and Polity, 9,* 61–75.

Graham, M., Straumann, R. K., & Hogan, B. (2015). Digital divisions of labor and informational magnetism: Mapping participation in Wikipedia. *Annals of the Association of American Geographers, 5608,* 1–21.

Grant-Muller, S. M., Gal-Tzur, A., Minkov, E., Kuflik, T., Nocera, S., & Shoor, I. (2015). Transport policy: Social media and user-generated content in a changing information paradigm. In S. Nepal & C. Paris (Eds.), *Social media for government services* (pp. 325–366). Springer.

Hall, G. B., Chipeniuk, R., Feick, R. D., Leahy, M. G., & Deparday, V. (2010). Community-based production of geographic information using open source software and Web 2.0. *International Journal of Geographical Information Science, 24*(5), 761–781.

Hardy, D. (2013). The geographic nature of Wikipedia authorship. In D. Sui, S. Elwood, & M. Goodchild (Eds.), *Crowdsourcing geographic knowledge: Volunteered geographic information (VGI) in theory and practice* (pp. 175–200). Springer.

Higgs, G., Berry, R., Kidner, D., & Langford, M. (2008). Using IT approaches to promote public participation in renewable energy planning: Prospects and challenges. *Land Use Policy, 25,* 596–607.

Hollenstein, L., & Purves, R. (2010). Exploring place through user-generated content: Using Flickr to describe city cores. *Journal of Spatial Information Science, 1,* 21–48.

Innes, J. E., & Booher, D. E. (2004). Reframing public participation: Strategies for the 21st century. *Planning Theory & Practice, 5,* 419–436.

Irvin, R. A., & Stansbury, J. (2004). Citizen participation in decision making: Is it worth the effort? *Public Administration Review, 64*(1), 55–65.

Johannessen, M. R., Flak, L. S., & Sæbø, Ø. (2012). Choosing the right medium for municipal eParticipation based on stakeholder expectations. In H. J. Scholl, M. Janssen, M. A. Wimmer, C. E. Moe, & L. S. Flak (Eds.), *Electronic government: 11th IFIP WG 8.5 International Conference, EGOV 2012* (pp. 25–36). Springer.

Johnson, P. A., & Sieber, R. E. (2013). Situating the adoption of VGI by government. In D. Sui, S. Elwood, & M. Goodchild (Eds.), *Crowdsourcing geographic knowledge: Volunteered geographic information (VGI) in theory and practice* (pp. 65–81). Springer.

Kergosien, E., Laval, B., Roche, M., & Teisseire, M. (2014). Are opinions expressed in land-use planning documents? *International Journal of Geographical Information Science, 28,* 739–762.

Khan, G. F., Yoon, H. Y., Kim, J., & Park, H. W. (2013). From e-government to social government: Twitter use by Korea's central government. *Online Information Review, 38,* 95–113.

Kim, K.-S., Kojima, I., & Ogawa, H. (2016). Discovery of local topics by using latent spatio-temporal relationships in geo-social media. *International Journal of Geographical Information Science, 8816,* 1–24.

King, C. S., Feltey, K. M., & Susel, B. O. (1998). The question of participation: Toward authentic public participation in public administration. *Public Administration Review, 58,* 317–326.

Lawrence, H., Robertson, C., Feick, R., & Nelson, T. (2015). Identifying optimal study areas and spatial aggregation units for point-based VGI from multiple sources. In F. Harvey & Y. Leung (Eds.), *Advances in spatial data handling and analysis* (pp. 65–84). Springer.

Leetaru, K. H., Wang, S., Cao, G., Padmanabhan, A., & Shook, E. (2013). Mapping the global Twitter heartbeat: The geography of Twitter. *First Monday, 18,* 1–12.

Lenhart, A., & Fox, S. (2009, February 12). *Twitter and status updating.*

Li, L., & Goodchild, M. F. (2012*). Constructing places from spatial footprints. In Proceedings of the 1st ACM SIGSPATIAL International Workshop on Crowdsourced and Volunteered Geographic Information—GEOCROWD 12* (pp. 15–21). ACM.

Lin, W. (2013). When Web 2.0 meets public participation GIS (PPGIS): VGI and spaces of participatory mapping in China. In D. Sui, S. Elwood, & M. Goodchild (Eds.), *Crowdsourcing geographic knowledge: Volunteered geographic information (VGI) in theory and practice* (pp. 83–103). Springer.

Lowndes, V., Pratchett, L., & Stoker, G. (2001). Trends in public participation: Part 1—local government perspectives. *Public Administration, 79,* 205–222.

Massa, P. (2015).*Social media geographic information (SMGI): Opportunities for spatial planning and governance* [Doctoral dissertation, Università degli Studi di Cagliari]. UNICA.

Massa, P., & Campagna, M. (2014). Social media geographic information: Recent findings and opportunities for smart spatial planning. *TeMA: Journal of Land Use, Mobility and Environment, 7*(Special Issue: INPUT 2014), 645–658.

McCall, M. K., & Dunn, C. E. (2012). Geo-information tools for participatory spatial planning: Fulfilling the criteria for "good" governance? *Geoforum, 43,* 81–94.

Mossberger, K., Wu, Y., & Crawford, J. (2013). Connecting citizens and local governments? Social media and interactivity in major U.S. cities. *Government Information Quarterly, 30*(4), 351–358.

Panagiotopoulos, P., Bigdeli, A. Z., & Sams, S. (2014). Citizen–government collaboration on social media: The case of Twitter in the 2011 riots in England. *Government Information Quarterly, 31,* 349–357.

Panteras, G., Wise, S., Lu, X., Croitor, A., Crooks, A., & Stefanidis, A. (2014). Triangulating social multimedia content for event localization using Flickr and Twitter. *Transactions in GIS, 19,* 694–715.

Pocewicz, A., Nielsen-Pincus, M., Brown, G., & Schnitzer, R. (2012). An evaluation of Internet versus paper-based methods for Public Participation Geographic Information Systems (PPGIS). *Transactions in GIS, 16,* 39–53.

Rattenbury, T., & Naaman, M. (2009). Methods for extracting place semantics from Flickr tags. *ACM Transactions on the Web, 3,* 1–30.

Reddick, C. G., & Norris, D. F. (2013). Social media adoption at the American grass roots: Web 2.0 or 1.5? *Government Information Quarterly, 30*(4), 498–507.

Region of Waterloo. (2007). *A blueprint for shaping growth in Waterloo Region.* Retrieved May 16, 2016 from http://www.regionofwaterloo.ca/en/ngBusiness/resources/BlueprintShapingGrowth.pdf

Ríos, S., & Muñoz, R. (2014). Content patterns in topic-based overlapping communities. *The Scientific World Journal, 2014,* 1–11.

Robertson, C., & Feick, R. (2016). Bumps and bruises in the digital skins of cities: unevenly distributed user-generated content across US urban areas. *Cartography and Geographic Information Science, 43*(4), 283–300. https://doi.org/10.1080/15230406.2015.1088801

Rowe, G. (2005). A typology of public engagement mechanisms. *Science, Technology and Human Values, 30,* 251–290.

Sandoval-Almazan, R., & Gil-Garcia, J. R. (2014). Towards cyberactivism 2.0? Understanding the use of social media and other information technologies for political activism and social movements. *Government Information Quarterly, 31*(3), 365–378.

Schweitzer, L. (2014). Planning and social media: A case study of public transit and stigma on Twitter. *Journal of the American Institute of Planners, 80,* 218–238.

Shelton, T., Poorthuis, A., & Zook, M. (2015). Social media and the city: Rethinking urban socio-spatial inequality using user-generated geographic information. *Landscape and Urban Planning, 142,* 198–211.

Sieber, R. (2006). Public participation geographic information systems: A literature review and framework. *Annals of the Association of American Geographers, 96,* 491–507.

Sieber, R. E., & Johnson, P. A. (2015). Civic open data at a crossroads: Dominant models and current challenges. *Government Information Quarterly, 32,* 308–315.

Silva, T. H., Vaz de Melo, P. O. S., Almeida, J. M., & Loureiro, A. A. F. (2013). Social media as a source of sensing to study city dynamics and urban social behavior: Approaches, models, and opportunities. In M. Atzmueller, S. Chin, D. N. Kanellopoulos, & T. Bansal (Eds.), *Analysis of social media for sentiment and topic extraction* (pp. 63–87). Springer.

Slotterback, C. S. (2011). Planners' perspectives on using technology in participatory processes. *Environment and Planning b: Planning and Design, 38,* 468–485.

Smith, A., Hawes, T., & Myers, M. (2014). Hierarchie: Interactive visualization for hierarchical topic models. *Association for Computational Linguistics, 2014,* 71–78.

Stefanidis, A., Crooks, A., & Radzikowski, J. (2013). Harvesting ambient geospatial information from social media feeds. *GeoJournal, 78,* 319–338.

Tao, K., Abel, F., Hauff, C., & Houben, G. (2012). What makes a tweet relevant for a topic? *CEUR Workshop Proceedings, 838,* 49–56.

Thelwall, M., Buckley, K., & Paltoglou, G. (2012). Sentiment strength detection for the social web. *Journal of the American Society for Information Science and Technology, 63,* 163–173.

Vicente, M. R., & Novo, A. (2014). An empirical analysis of e-participation: The role of social networks and e-government over citizens' online engagement. *Government Information Quarterly, 31*(3), 379–387.

Wang, W., & Stewart, K. (2015). Spatiotemporal and semantic information extraction from Web news reports about natural hazards. *Computers, Environment and Urban Systems, 50,* 30–40.

Wukich, C., & Mergel, I. (2016). Reusing social media information in government. *Government Information Quarterly, 33*(2), 384–391.

Yin, J., Lampert, A., Cameron, M., Robinson, B., & Power, R. (2012). Using social media to enhance emergency situation awareness. *IEEE Intelligent Systems, 27,* 52–59.

Zavattaro, S. M., French, P. E., & Mohanty, S. D. (2015). A sentiment analysis of U.S. local government tweets: The connection between tone and citizen involvement. *Government Information Quarterly, 32*(3), 333–341.

Zhai, C., & Lafferty, J. (2001). A study of smoothing methods for language models applied to ad hoc information retrieval. In Proceedings of the 24th Annual International ACM SIGIR *Conference on Research and Development in Information Retrieval* (pp. 334–342). ACM.

Chapter 7
Conclusion

Abstract This chapter concludes the book by summarizing key findings on the conceptualization and practices of geo-participation. Conceptually, the proposed typology classifies geo-participation practices by differentiating the tools, methods, and outcomes of geo-participation and identifies opportunities and challenges of researching and implementing geo-participation. Empirically, this book tells a story of opportunities and challenges that sheds light on how local governments can use geosocial media as: (1) a data source for soliciting citizen input and (2) a tool for sharing and reporting issues related to routine government services and enabling new forms of government–citizen interaction.

7.1 Empirical Findings and Contributions: Revisiting Research Questions

This book puts forward a holistic view of how advances in geospatial technologies have given rise to evolving government–citizen relationships. Geo-participation practices are classified into three types, with enabling technologies, data, and participation outcomes of each type of geo-participation identified. Opportunities for local governments to solicit public input and collaborate with citizens using emerging geo-participation practices are illustrated. The typology is then applied to both academic literature and government programs to demonstrate the use of the typology. The classification of academic literature suggests that different types of geo-participation are associated with varying research focuses. In practice, different types of geo-participation are associated with different participation processes and outcomes.

The empirical studies further examine the potential of the emerging geo-participation practices. Chapters 4 and 5 examine the transactional participation practices using the case studies of citizen reporting non-emergency issues through geosocial media. Citizen survey and local government interviews suggest some common data quality issues related to biased representation of geosocial media data, particularly the uneven spatial distribution and skewed representation of data contributors.

S. Zhang, *Geo-participation*, https://doi.org/10.1007/978-981-95-8228-0_7

One interesting finding of the study is that citizens consider privacy issues differently according to the context of data usage. Previous studies suggest that concerns of privacy may prevent people from contributing location data and result in scarcity or incompleteness of contributed data (Elwood & Leszczynski, 2013). This study suggests that citizen participants have fewer privacy concerns for providing location information when they know the use of the information. This finding has implications for alleviating participants' privacy concerns and encouraging participation by providing participants with details of what contributed data are used for and how.

Another significant finding of this study is that government staff and citizens have contradictory views on spatial data quality of geosocial media, which have implications for the opportunities and challenges of local government adopting geosocial media for municipal reporting. Local governments face significant challenges of adopting geosocial media considering its poor quality, as they need to invest extra time and human capital to process and validate geosocial media reports. Accordingly, local governments are less likely to meet citizen expectations of having an immediate response from local governments. Developing integrated and automated methods is necessary for local governments to make efficient uses of geosocial media reporting and to improve their operational efficiencies of handling citizen requests in the long term. In practice, local governments should determine the proper practice of adopting geosocial media reporting based on their needs and current capacities. Models of implementing geosocial media reporting suggested in Chap. 5 have practical contributions to government operations, as it provides local government with a reference of implementing geosocial media reporting for municipal reporting programs. Citizens, on the other hand, find geosocial media reports sufficiently accurate for them to understand dynamic urban environments. This suggests the potential of developing collaborative civic actions of citizens sharing non-emergency issues outside the mandate of governments. Resonating with "Do-it-Yourself" model as suggested by Linders (2012), citizens may develop their own public services and use contributed geospatial data based on their own information needs. From the data perspective, differences in how citizens and local government staff perceive the quality of geosocial data present the need to address users' information needs when evaluating the quality of VGI. While data quality is considered important to VGI studies, how perceived qualities of VGI may differ among different groups of users has not been well addressed in the current literature.

Chapter 6 further investigates the usefulness of geosocial media as a data source for understanding public needs. The empirical study suggests that geosocial media can provide topic- and location-specific types of public input that differ from what might be found in complementary data sources. Particularly, some of the public input collected from geosocial media is relevant to physical infrastructure and quality of public services. This input is akin to citizen requests for public service, which are usually handled by customer-service programs of municipalities. The finer temporal resolution of geosocial media data provides governments with opportunities to probe into changes in public perceptions and to monitor dynamic urban environments.

From a geospatial data perspective, the study demonstrates that the use of geosocial data is most likely to be limited by spatial, temporal, and semantic relevance.

Contesting the notion that VGI studies should primarily focus on the volume of information (Brown, 2017), data that are relevant to a particular study may only be a small subset of the big raw data. As suggested by Poorthuis and Zook (2017), social and urban researchers may often need to carefully extract a small subset of the big data and obtain in-depth insights on social and spatial phenomena using the small data. As such, analyzing geosocial data often requires a mixture of quantitative and qualitative methods (Shelton et al., 2014). That is, quantitative methods can help extract, categorize, and understand the general patterns of the data, whereas qualitative interpretation is necessary to obtain deeper insights from the data.

7.2 Implications for Local Governments: Geospatial Data as Conduits for Government–Citizen Collaboration

This book suggests that significant opportunities have emerged from mass geospatial data generation for local governments to better address public needs and enhance government–citizen collaboration. First, this research demonstrates the potential for local government to use data-driven approaches to solicit public opinions from passively generated geosocial data. While the concept of citizen sensing is not new, its application in local government decision-making has not been widely studied. Further integration of citizen sensing with routine government service may help local governments improve service response time and deliver services that are better tailored to citizen needs (Lee & Kwak, 2012; Linders et al., 2015). Second, this research suggests that collective data contribution can serve as a conduit for facilitating public participation and spurring innovative forms of collaboration. Chapter 3 suggests the potential of citizens collaboratively sharing information in supplement to government services. Chapters 4 and 5 further assesses this potential from a geospatial data perspective and validates the usefulness of geosocial data for citizens to understand dynamic environments. Potentially, geospatial information may not only be transmitted between citizens and governments, but also be shared among citizens so that citizens can use the information for their own needs.

However, local governments face significant challenges to take advantage of these emerging opportunities. Tensions arise from increased citizen demands and limited local-government resources to handle citizen input. In particular, the uncertain quality and accordingly high costs of handling citizen requests will likely be a continuing threat to local government adoption of contributed data. Local governments do not always have the capacity (e.g., financial and human resources, technical capacities) to capitalize on technical developments including adopting new tools and methods of processing citizen requests.

In practice, the process of diffusing technological innovation is affected not only by technological developments but also by other organizational and administrative factors (Mustonen-Ollila & Lyytinen, 2003). According to the well-established theory of diffusion of innovation, perceived advantage or improvement of using a new idea, approach or system over existing ones will motivate individuals or organizations to adopt innovations (Rogers, 1983). Yet, adopters will also consider other characteristics of an innovation including compatibility, trialability, complexity, and observability (Rogers, 1983).

In the government context, compatibility refers to how well an innovation aligns with existing procedures and norms in the organization (Greenhalgh et al., 2004). Trialability refers to how easy the organization can test the innovation and assess the impacts of innovations on the organization (Lundblad, 2003). According to Weber (1947), operations in a bureaucratic organization such as governments need to follow "a set of formal, explicit, comprehensive and stable rules that are impersonally enforced in decision-making and lead to predictable and determinate results" (Cordella & Tempini, 2015, p. 280). Therefore, governments usually carry out repeated and deliberative processes to ensure the smooth transition of government operations and to minimize the possibility of having negative consequences (Bertot et al., 2016).

Janowski (2015) suggested that innovation in governments would usually encompass several steps (Fig. 7.1). Governments often seek to adopt new digital technologies when facing economic, social, or political challenges. While the new technology may only be applied in response to a short-term pressure at first, later it will be applied and improved repeatedly before it becomes a standard practice embedded in government operations. Eventually, the institutionalized practice should follow the bureaucratic legislative, policy, and administrative principles, and enforce "democratic values of impartiality, fairness and equality" (Cordella & Tempini, 2015, p. 280). Making a digital innovation become a standard government practice should therefore take holistic approaches that account for a variety of technological, administrative, policy, and institutional factors.

Complexity and observability of an innovation are also important to determining the adoption rate of the innovation (Rogers, 1983). Complexity refers to how easily adopters can understand and use an innovation. Observability refers to how visible an innovation is to adopters. Influenced by these factors, the extent to which an organization is willing to adopt an innovation and to change its familiar practices varies (Sahin, 2006). Rogers (1983) categorized adopters into innovators, early adopters, early majority, late majority, and laggards. Often, innovators and early adopters would take the leadership roles in adopting new ideas and in providing information and advice about the innovation (Sahin, 2006). Early majority often follow early adopters but do not take the leadership role in the innovation diffusion process. Late

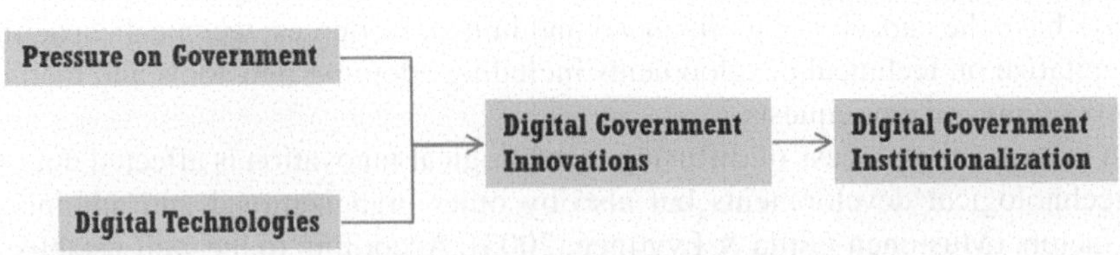

Fig. 7.1 Digital government innovation flow. Adapted from Janowski (2015)

majority and laggards usually are more skeptical about the innovation and its associated outcomes. They tend to decide whether to adopt the innovation after they see successful adoption examples from other adopters. In accordance with Rogers' categorization, the interviews of Chap. 4 showed that the adoption rates of municipalities vary significantly. Some municipalities have integrated geosocial media reporting with their citizen reporting systems, while others have not. Some of the interviewed municipalities that have not used geosocial media reporting expressed the desire to learn from others' experiences to determine whether or not they would adopt geosocial media reporting. According to Rogers (1983), networking among adopters may help reduce the uncertainty of innovation adoption and help late adopters to feel safer about adopting the innovation. It is therefore helpful for cities that led technological changes to share their experiences of implementing technology-driven initiatives so that other cities could better understand the processes and outcomes of implementing an innovation (Johnson & Sieber, 2013).

7.3 Implications for Geo-Participation Research

The implications of this study for geo-participation research are twofold. From a data perspective, an important thread throughout the empirical studies is that the participatory paradigm of spatial data generation may give rise to a deluge of unstructured, georeferenced data that require new validating and interpreting methods (Miller & Goodchild, 2015). Since geosocial data are rich in content and unstructured, qualitative data analysis methods (e.g., text-analytics) and standardized frameworks for managing geosocial data are necessary. The TFA toolkit presented in Chap. 6 makes a methodological contribution to automating the processing of unstructured qualitative citizen data (Brooker, Barnett, & Cribbin, 2016). Research should also further develop conceptual understandings of the data (Kitchin, 2014a). Much of the citizen-generated geospatial data relate to their perceptions and experiences with local environments that are subject to time and people (Roche & Feick, 2012). According to Goodchild (2011), space-based conceptualization and methods are often limited in representing and analyzing these data with vague and contextual information; the concept of place should be used instead to conceptualize implicit geographic information that is rich in everyday lived experiences. Other ethical (e.g., privacy) and scientific challenges (e.g., idiographic versus nomothetic knowledge; sampling of the data) should also be further studied.

From a participation perspective, this book advances the understanding of geo-participation by classifying geo-participation practices and identifying the application of each type of geo-participation in local government settings. The empirical studies examined two participation practices (passive and transactional geo-participation). From the viewpoint of tool development, these emerging participation methods generate needs for new tools for data analysis that support the specific needs of government officials (e.g., extracting exact locations from ambiguous place descriptions, identifying height and parcel information for non-emergency reporting). From the viewpoint of implementing public participation, this book suggests that practices such as passive VGI and citizen sensing can be used for participatory uses. While some scholars argue that VGI is not participatory (Brown, 2017; McCall et al., 2013), this study suggests that the practice of citizen generating geospatial data and the corresponding data outcomes can be repurposed to help local government with decision-making and facilitate government–citizen collaboration. In particular, the work presented in this book demonstrates the potential for local governments to enhance their capabilities of soliciting public input, which lay the foundation for governments to promote open participation by utilizing the public input (Lee & Kwak, 2012) and to develop citizen-centric services that are better tailored to citizen needs (Reddick et al., 2016). However, neither geosocial media nor passive VGI in general could be stand-alone approaches for collecting public input. Similar to findings from other work (Zook, 2017), the study suggests the limitations of geosocial data in terms of their biased representation, as public input collected from geosocial media and user letters reveal different interests and concerns of the public. It is necessary to develop integrated participatory approaches that can elicit public input from different channels and obtain comprehensive views of public needs.

7.4 Ethical Issues of Passive Geo-Participation

There are also several ethical issues associated with passive geo-participation that need to be further addressed both in research and in practice. Kitchin (2016) suggested that the ubiquitous data collection through sensors, the Internet, telecommunication, and GPS devices has posited several privacy concerns. That is, information about people's identity, communication, transaction, location, and movement might be accessed, shared, and used without people's knowledge or consent. The challenge is that people are often not aware of what data are gathered and how the data will be used, and therefore do not have the opportunity to consent or refuse the data collection and use (Crawford & Finn, 2015). Solove (2013) argued that the design of web and mobile applications often provides users with little power to bargain regarding their privacy. That is, users often have no choice but to consent with the application's user terms in order to use the application. As a result, the data may be collected regardless of users' preferences. Moreover, the data or the collection of data may include personally identifying information (Kar & Ghose, 2014). Montjoye, Radaelli, and Singh (2012) showed an example that customers' personal information could become identifiable from anonymous transactional records when the records are combined with Instagram photos. The misuse of this information may raise unintended consequences such as safety concerns (Olteanu et al., 2016).

Context is important to the understanding of people's privacy preferences. Crawford and Finn (2015) pointed out that people's choices of whether to disclose personal information or not may shift according to the circumstances. As shown in Chap. 4, people are more willing to disclose their locational information when they know the information is used for specific purposes. This result has indications that people may have fewer privacy concerns when perceived benefits outweigh risks (Ricker et al., 2015). However, we should be cautious about arbitrary and sometimes falsified gauges of perceived benefits and risks. For example, Tene and Polonetsky (2012) stated that "where the benefits of prospective data use clearly outweigh privacy risks, the legitimacy of processing should be assumed even if individuals decline to consent" (p. 67). Crawford and Finn (2015) criticized such statement by asking "who gets to decide when the benefits outweigh the risks?" (p. 499).

There is considerable skepticism about the legitimacy of using citizen-contributed data if the decisions of using the data are not determined by the communities or individuals who contribute the data. In the context of crisis management, Crawford and Finn (2015) suggested that although mining information from data sources such as geosocial media is beneficial for understanding and assessing disaster situations, it is inappropriate to give less consideration to privacy and ethical considerations because of these benefits. Rather, the benefits and risks of a data use should be carefully deliberated by considering the context of data uses and the preferences of data contributors.

With regard to location privacy, methodological developments may also assist with protecting users' privacy. Kar and Ghose (2014) suggested four computational approaches for protecting users' location privacy, including anonymity, aggregation, obfuscation, and encryption. Anonymity refers to making a user's location anonymous by using one's location that is shared with other users. Personal location can also become unidentifiable by aggregating individual locations to groups. Obfuscation refers to methods of introducing noise to locational data so that the accuracy of data will be reduced. A true location may also be encrypted or fudged so that it cannot be identified. While the technical details and applications of these computational approaches are beyond the scope of this discussion, these methods shed light on the use of users' locational data. For instance, depending on the characteristics of a locational dataset, certain computational approaches may be applied to the dataset to ensure that personal location information is not disclosed.

Passive geo-participation builds upon the assumption that collecting dynamic and fine-grained data about aspects of urban life could inform evidence-based decisions and enable effective modes of governance (Kitchin, 2014b). It is necessary to further consider and improve regulations on dissemination and application of data. Kitchin (2014a, b) warned us that the lack of regulated enforcement may raise significant resistance of data-driven city governance concerning the abuses of data. Citizen-generated data and other forms of sensor data collection may raise new challenges to data-related ethical and legal issues concerning intellectual property, liability, privacy, and licensing (Scassa, 2013). For instance, the recent openings of real-time trajectory data introduce complexity to the issue of data licensing given that the licensing of such dynamic data is not consistent with the licensing of static data (Scassa & Diebel, 2017). Moreover, inconsistent regulations among different levels of governments further complicate the issue of data regulation (Scassa & Diebel, 2017). That is, while the use of a dataset meets the requirement of one level of government, it may not meet the requirement of another.

Attention should also be paid to inherent challenges associated with data-driven and algorithm-based approaches (Janssen & Kuk, 2016). For example, algorithms shape how data are collected, processed, and interpreted (Kwan, 2016). For instance, Twitter offers different levels of access to retrieving user feeds through its APIs. What tweets can be harvested from the Twitter API is influenced by Twitter's algorithms (Joseph et al., 2014). Using different data collection methods (e.g., adopting geosocial media collection software vs. "ad hoc" data collection via the open-source API libraries) may also result in different datasets (Poorthuis & Zook, 2017). The resulting dataset can only represent a selective sample of whole Twitter users, yet how the sample is selected is often left unknown. How data are analyzed and interpreted also have significant impacts on what insights are obtained from the data. The widely cited paper by Lazer et al. (2014) pointed out the errors in flu prediction based on Google flu trend (i.e., the big data overfitted the small number of cases) and warned us of the trap in big data analysis. That is, ad hoc analysis methods are often adopted without careful model calibrations and evaluations of the replicability of data analysis.

The conclusions drawn from data analysis are also subject to the assumptions made for the analysis. For instance, a range of studies have used geosocial media for analyzing human mobility patterns and people's preferences for using city places (Chen et al., 2017; Liu et al., 2014). These studies provided city planners with evidence on measuring the quality of city design and services for satisfying citizens' needs of transportation, public infrastructure, and services. Such evidence should be used cautiously, however, as it may reflect only one artifact of city life (Kitchin, 2014b). For example, how people use city spaces might be constrained by how the city is currently designed and other cultural and policy factors and may not always be equivalent to how people would like to use city spaces. The muddle of the two may overlook factors that should be accounted for in city planning.

Practitioners therefore should be attentive to the control possessed by data and algorithms over what information is included or excluded and over what conclusions are drawn (Janssen & Kuk, 2016). Particularly, attention should be paid to what questions can be answered by the data, the extent to which the algorithms impact the representation of different social groups, if and how different algorithms may result in different conclusions, and the validity and replicability of algorithms that are used for collecting and analyzing the data (Kwan, 2016). Failure to address these questions in the context of city governance may lead to improper understandings of public perceptions (e.g., the over- or under-representation of certain social groups) or reinforcement of undesired practices (Janssen & Kuk, 2016). As suggested by Vayena et al. (2015), ensuring the robustness of algorithms for data analysis should not only be scientific inquiries but should also be ethical requirements.

Algorithm-based approaches for governance also raise concerns about the transparency and accountability of governments. Algorithms behind how data are processed and used have become increasingly autonomous and invisible (Fink, 2017). Communicating complex algorithms is often challenging, and only a few experts may understand algorithms for data analysis (Sandvig, 2015). Accordingly, the lack of transparency and complexity may deprive the public's capabilities to understand and scrutinize decisions made by the governments (Coletta & Kitchin, 2017). Janssen and

Kuk (2016) suggest that this challenge indicates higher requirements on decision-makers' skills and expertise to ensure the validity of algorithms used for making decisions. Issues that shape and arise from the use of algorithms in governance should also be further explored both in research and in practice (Coletta & Kitchin, 2017).

7.5 Limitations and Future Research

While this book makes substantial contributions to understanding geo-participation as a thread of current government efforts to enhance public participation, it demonstrates a few shortcomings. First, the empirical study presented in Chap. 5 is limited by who participated in the citizen survey. The study sample of citizen participants tends to overrepresent people who are aged between 25 and 35. Other studies have characterized the majority of social media users as young, well-educated, and more affluent and suggested the potential of social media as a channel for engaging the young population who are often underrepresented in participatory projects (Schweitzer, 2014). Yet, this potential needs to be further validated using qualitative data that provide evidence on the representativeness of geosocial data. This is also important for addressing the potential of using data-driven approaches to mine public opinions as presented in Chap. 6. It is necessary to further assess whether the presented approach, or more broadly automatic citizen sensing from various sources, can address issues of citizen participation including low participation rates and underrepresentation of certain population groups.

Second, while this book primarily focuses on technological and organizational aspects of implementing geo-participation, the social processes embedded in geo-participation require further attention. Chapter 3 points to the need to reconsider the notion of "local" in citizen participation. This issue should be further investigated by identifying how ICTs change social and power relationships among governments and citizens and what implications these changing relationships have on digital divide, social disparity and citizen participation (Haworth, 2017).

Moreover, empirical studies in the book provide in-depth examinations of geosocial media, which is only one popular channel of citizens contributing geospatial data. It would be advantageous to examine citizen contributions from other ICT-based channels and in other contexts. I envision the following key areas for further scientific investigation.

First, it would be beneficial to carry out comparative studies to investigate the similarities and differences among multiple ICT-based channels. Chapter 6 demonstrates differences in public input elicited from different channels. Further investigation along this line may shed light on if and how different participatory approaches may supplement each other and whether an integrated participatory approach could be developed accordingly. Moreover, with continuing developments of web technologies and Internet of Things (IoT), it would be interesting to examine the extent to which automated citizen sensing from various physical and human sensors may help solicit public input or even substitute methods of purposively collecting public input from human participants. Additionally, Bright and Margetts (2016) raised the issue that since people are passively contributing data, they may not expect their opinions to be integrated into decision-making processes. This contrasts with traditional public participation approaches, where people consciously give voice and make choices. It is necessary to further investigate how the conceptualization of participation may change accordingly.

Second, with the increasing heterogeneous geospatial data generated by the public, spatial data quality must be addressed from a multi-user perspective. While this study suggests different perspectives of users (i.e., citizens and governments) toward spatial data quality, it is necessary to examine how perspectives of different citizen groups vary with respect to their needs. Such investigation is necessary for developing customized public services, similar to the proof-of-concept service of citizens collaboratively contributing sidewalk information. For example, how do people perceive accessibility differently? Accordingly, how should spatial data quality of sidewalk information be evaluated differently? Future studies should focus on developing spatial data quality measurements that are tailored to users' needs.

Lastly, recent developments in Large Language Models (LLMs) have provided new opportunities in assisting public participation. The abilities in understanding and generating natural language content allow LLMs to simulate public opinions that are comparable to human beings (Zhou et al., 2024a, b). Future work may further examine the potential of LLMs as a new means of public participation and how it may add to existing conceptualization of geo-participation. While having similarities with data-driven passive participation, LLMs-driven participation may be critically assessed by accounting for various technological, organizational, and ethical issues.

References

Bertot, J. C., Estevez, E., & Janowski, T. (2016). Universal and contextualized public services: Digital public service innovation framework. *Government Information Quarterly, 33*(2), 211–222. https://doi.org/10.1016/j.giq.2016.05.004

Bright, J., & Margetts, H. (2016). Big data and public policy: Can it succeed where E-participation has failed? *Policy and Internet, 8*(3), 218–224.

Brooker, P., Barnett, J., & Cribbin, T. (2016). Doing social media analytics, (December), 1–12. https://doi.org/10.1177/2053951716658060

Brown, G. (2017). A review of sampling effects and response bias in Internet participatory mapping (PPGIS/PGIs/VGI). *Transactions in GIS, 21*(1), 39–56. https://doi.org/10.1111/tgis.12207

Chen, Y., Liu, X., Li, X., Liu, X., Yao, Y., Hu, G., Xu, X., & Pei, F. (2017). Delineating urban functional areas with building-level social media data: A dynamic time warping (DTW) distance based k-medoids method. *Landscape and Urban Planning, 160,* 48–60.

Coletta, C., & Kitchin, R. (2017). Algorhythmic governance: Regulating the "heartbeat" of a city using the Internet of Things. *Big Data & Society, 4*(2), 205395171774241. https://doi.org/10.1177/2053951717742418

Cordella, A., & Tempini, N. (2015). E-government and organizational change: Reappraising the role of ICT and bureaucracy in public service delivery. *Government Information Quarterly, 32*(3), 279–286. https://doi.org/10.1016/j.giq.2015.03.005

Crawford, K., & Finn, M. (2015). The limits of crisis data: Analytical and ethical challenges of using social and mobile data to understand disasters. *GeoJournal, 80*(4), 491–502. https://doi.org/10.1007/s10708-014-9597-z

Elwood, S., & Leszczynski, A. (2013). New spatial media, new knowledge politics. *Transactions of the Institute of British Geographers, 38*(4), 544–559. https://doi.org/10.1111/j.1475-5661.2012.00543.x

Fink, K. (2017). Opening the government's black boxes: Freedom of information and algorithmic accountability. *Information Communication and Society, 4462*(May), 1–19. https://doi.org/10.1080/1369118X.2017.1330418

Goodchild, M. F. (2011). Formalizing place in geographic information systems. In: Burton, L., Matthews, S., Leung, M., Kemp, S., Takeuchi, D. (eds) *Communities, Neighborhoods, and Health. Social Disparities in Health and Health Care*, vol 1. Springer, New York, NY. https://doi.org/10.1007/978-1-4419-7482-2_2

Greenhalgh, T., Robert, G., Macfarlane, F., Bate, P., & Kyriakidou, O. (2004). Diffusion of innovations in service organizations: Systematic review and recommendations. *The Milbank Quarterly, 82*(4), 581–629. https://doi.org/10.1111/j.0887-378X.2004.00325.x

Haworth, B. (2017). Implications of volunteered geographic information for Disaster Management and GIScience: A more complex world of volunteered geography. *Annals of the American Association of Geographers, 4452*(July), 1–15. https://doi.org/10.1080/24694452.2017.1321979

Janowski, T. (2015). Digital government evolution: From transformation to contextualization. *Government Information Quarterly, 32*(3), 221–236. https://doi.org/10.1016/j.giq.2015.07.001

Janssen, M., & Kuk, G. (2016). The challenges and limits of big data algorithms in technocratic governance. *Government Information Quarterly, 33*(3), 371–377. https://doi.org/10.1016/j.giq.2016.08.011

Johnson, P. A., & Sieber, R. E. (2013). Situating the adoption of VGI by government. In D. Sui, S. Elwood, & M. Goodchild (Eds.), *Crowdsourcing geographic knowledge: Volunteered geographic information (VGI) in theory and practice* (pp. 65–81). Springer.

Joseph, K., Landwehr, P. M., & Carley, K. M. (2014, April). Two 1% s don't make a whole: Comparing simultaneous samples from Twitter's streaming API. In *International conference on social computing, behavioral-cultural modeling, and prediction* (pp. 75–83). Cham: Springer International Publishing.

Kar, B., & Ghose, R. (2014). Is my information private? Geo-privacy in the World of Social Media. *CEUR Workshop Proceedings*, 1273(Webster), 28–31.

Kitchin, R. (2014a). The data revolution. *SAGE Publications Ltd*. https://doi.org/10.1017/CBO9781107415324.004

Kitchin, R. (2014b). The real-time city? Big data and smart urbanism. *GeoJournal*, (June). Retrieved from http://link.springer.com/article/10.1007/s10708-013-9516-8

Kitchin, R. (2016). *Getting smarter about smart cities: Improving data privacy and data security*. Data Protection Commissioner of Ireland.

Kwan, M. P. (2016). Algorithmic geographies: Big data, algorithmic uncertainty, and the production of geographic knowledge. *Annals of the American Association of Geographers, 106*(2), 274–282. https://doi.org/10.1080/00045608.2015.1117937

Lazer, D., Kennedy, R., King, G., & Vespignani, A. (2014). The Parable of Google Flu: Traps in Big Data Analysis. *Science, 343,* 1203–1205. https://doi.org/10.1126/science.1248506

Lee, G., & Kwak, Y. H. (2012). An Open Government Maturity Model for social media-based public engagement. *Government Information Quarterly, 29*(4), 492–503. https://doi.org/10.1016/j.giq.2012.06.001

Linders, D. (2012). From e-government to we-government: Defining a typology for citizen coproduction in the age of social media. *Government Information Quarterly, 29*(4), 446–454. https://doi.org/10.1016/j.giq.2012.06.003

Linders, D., Liao, C. Z., & Wang, C. (2015). Proactive e-Governance: Flipping the service delivery model from pull to push in Taiwan. *Government Information Quarterly*. https://doi.org/10.1016/j.giq.2015.08.004

Liu, Y., Sui, Z., Kang, C., & Gao, Y. (2014). Uncovering patterns of inter-urban trip and spatial interaction from social media check-in data. *PLoS ONE, 9*(1), Article e86026.

Lundblad, J. P. (2003). A review and critique of Rogers â€tm diffusion of innovation theory as it applies to organizations. *Organization Development Journal, 21*(4), 50–64.

McCall, M. K., Martinez, J., & Verplanke, J. (2013). Shifting boundaries of volunteered Geographic in-formation systems and modalities: learning from PGIs. *An International E-Journal for Critical Geographies,* 1–36.

Miller, H. J., & Goodchild, M. F. (2015). Data-driven geography. *GeoJournal,* (October 2014), 449–461. https://doi.org/10.1007/s10708-014-9602-6

Montjoye, Y. De, Radaelli, L., & Singh, V. K. (2012). *Unique in the shopping mall: On the reidentifiability of credit card metadata.* Identity and Privacy.

Mustonen-Ollila, E., & Lyytinen, K. (2003). Why organizations adopt information system process innovations: A longitudinal study using Diffusion of Innovation theory. *Journal of Information Systems, 13,* 275–297. https://doi.org/10.1046/j.1365-2575.2003.00141.x

Olteanu, A., Castillo, C., Diaz, F., & Kiciman, E. (2016). Social Data: Biases, Methodological Pitfalls, and Ethical Boundaries. *SSRN Electronic Journal,* 1–47. https://doi.org/10.2139/ssrn.2886526

Poorthuis, A., & Zook, M. (2017). Making big data small: Strategies to expand urban and geographical research using social media. *Journal of Urban Technology, 24*(4), 115–135. https://doi.org/10.1080/10630732.2017.1335153

Reddick, C. G., Chatfield, A. T., & Ojo, A. (2016). A social media text analytics framework for double-loop learning for citizen-centric public services: A case study of a local government Facebook use. *Government Information Quarterly, 34,* 1–16. https://doi.org/10.1016/j.giq.2016.11.001

Ricker, B., Schuurman, N., & Kessler, F. (2015). Implications of smartphone usage on privacy and spatial cognition: Academic literature and public perceptions. *GeoJournal, 80*(5), 637–652.

Roche, S., & Feick, R. (2012). Wiki-place: Building place-based GIS from VGI, (2007), 2007–2009.

Rogers, E. M. (1983). *Diffusion of innovations* (3rd ed.). Free Press. http://books.google.com.sg/books?id=zw0-AAAAIAAJ&hl=en

Sahin, I. (2006). Detailed review of Rogers' Diffusion of Innovations theory and educational technology: Related studies based on Rogers' theory. *The Turkish Online Journal of Educational Technology, 5*(2), 14–23. https://doi.org/10.1287/mnsc.43.7.934

Sandvig, C. (2015). Seeing the sort: The aesthetic and industrial defense of "the algorithm." *Media-N: Journal of the New Media Caucus, 10*(3), 35–51.

Scassa, T. (2013). Legal issues with volunteered geographic information. *Canadian Geographer, 57*(1), 1–10. https://doi.org/10.1111/j.1541-0064.2012.00444.x

Scassa, T., & Diebel, A. (2016). Open or closed? Open licensing of real-time public sector transit data. *JeDEM—eJournal of eDemocracy and Open Government, 8*(2), 1–20.

Schweitzer, L. (2014). Planning and social media: A case study of public transit and stigma on Twitter. *Journal of the American Planning Association, 80*(3), 218–238. https://doi.org/10.1080/01944363.2014.980439

Shelton, T., Poorthuis, A., Graham, M., & Zook, M. (2014). Mapping the data shadows of Hurricane Sandy: Uncovering the sociospatial dimensions of "big data." *Geoforum, 52,* 167–179. https://doi.org/10.1016/j.geoforum.2014.01.006

Solove, D. J. (2013). Introduction: Privacy self-management and the consent dilemma. *Harvard Law Review, 126*(7), 1880–1903. https://doi.org/10.1525/sp.2007.54.1.23

Tene, O., & Polonetsky, J. (2012). Privacy in the age of big data: A time for big decisions. *Stanford Law Review Online, 64,* 63. https://doi.org/10.5121/ijgca.2012.3203

Vayena, E., Salathé, M., Madoff, L. C., & Brownstein, J. S. (2015). Ethical challenges of big data in public health. *PLoS Computational Biology, 11*(2), 1–7.

Weber, M. (1947). *The theory of social and economic organization*. Book (Vol. 57). https://doi.org/10.2307/2181723

Zhou, Z., Lin, Y., Jin, D., & Li, Y. (2024a). *Large language model for participatory urban planning* (No. arXiv:2402.17161). arXiv. https://doi.org/10.48550/arXiv.2402.17161

Zhou, Z., Lin, Y., & Li, Y. (2024b). *Large language model empowered participatory urban planning* (No. arXiv:2402.01698). arXiv. https://doi.org/10.48550/arXiv.2402.01698

Zook, M. (2017). Crowd-sourcing the smart city: Using big geosocial media metrics in urban governance. *Big Data & Society, 4*(1), 1–13. https://doi.org/10.1177/2053951717694384

The manufacturer's authorised representative in the EU is Springer Nature Customer Service Centre GmbH, Europaplatz 3, 69115 Heidelberg, Germany. If you have any concerns regarding our products, please contact ProductSafety@springernature.com

Printed and bound by CPI Group (UK) Ltd, Croydon, CR0 4YY
21/09/2026
02219968-0011